RESOURCES IN TECHNICAL COMMUNICATION

Outcomes and Approaches

W0234707

Edited by
Cynthia L. Selfe
The Ohio State University

Baywood's Technical Communications Series
Series Editor: CHARLES H. SIDES

Routledge
Taylor & Francis Group

LONDON AND NEW YORK

First published 2007 by Baywood Publishing Company, Inc.

2 Park Square, Milton Park, Abingdon, Oxon OX14 4RN
711 Third Avenue, New York, NY 10017, USA

Routledge is an imprint of the Taylor & Francis Group, an informa business

First issued in paperback 2017

Copyright © 2007 by Taylor & Francis

Library of Congress Catalog Number: 2006032957
ISBN 13: 978-0-89503-374-1 (hbk)

Library of Congress Cataloging-in-Publication Data

Resources in technical communication : outcomes and approaches / edited by Cynthia L. Selfe.
    p. cm. -- (Baywood's technical communications series)
  Includes bibliographical references and index.
  ISBN-13: 978-0-89503-374-1 (cloth : alk. paper)
  ISBN-10: 0-89503-374-7 (cloth : alk. paper) 1. Communication of technical information--Study and teaching. 2. Communication and technology. I. Selfe, Cynthia L., 1951-

  T10.5.R483 2007
  601--dc22

                         2006032957

ISBN 978-0-89503-374-1 (hbk)
ISBN 978-0-415-44047-9 (pbk)

Table of Contents

SECTION 1:
Focus on Rhetorical Understanding

Outcome: Students will understand the value of written communications as a problem-solving activity.

Outcome: Students will conceptualize reports as collections of rhetorical practices and improvisational strategies. Given a specific report template or format, they will construct a document that respects that template while also respecting their material and their rhetorical purposes.

Outcome: Students will understand the value of conducting both primary, experientially-based research and secondary-source research as a means for constructing rhetorically situated, persuasive communication.

Outcome: Students will understand how and why to use design approaches in developing technical and scientific communications and will be able to use different modalities rhetorically in developing communications.

SECTION 2:
Focus on Sociocultural Understanding

CHAPTER 10

Analyzing the Interactive Audience: Constructing a Communal
Knowledge Base

> **Outcome:** Students will understand how audiences interact with
> businesses within new digital contexts, and how they construct
> communal knowledge bases within such spaces.

SECTION 3:
Focus on the Complexities of Practice

CHAPTER 11

Understanding Usability Approaches

> **Outcome:** Students will develop a holistic understanding of
> usability, one that scaffolds common approaches according to
> their social complexities.

CHAPTER 12

An Ethics Primer: Strategies for Ethical Decision Making

> **Outcome:** Students will acquire a basic vocabulary and
> methodology for making ethical decisions in their technical
> communication practices.

CHAPTER 13

Select, Interpret, Produce: A Three-Part Model for Teaching
Information Graphics

> **Outcome:** Students will critically and competently select,
> interpret, and produce information graphics for use in technical
> documents.

CHAPTER 14

Listen Up! Oral Presentations in the Technical Communication
Classroom

> **Outcome:** Students will be able to author and deliver effective
> oral presentations that make appropriate and rhetorically effective
> use of available means, media, and modalities. Students will
> develop effective strategies for listening to and evaluating oral
> presentations.

> **Outcome:** Students will be able to speak and listen in small,
> informal group settings and collaborate effectively when
> presenting material.

> **Outcome:** Students will produce better, more polished, more
> refined prose; understand the different levels of editing and
> approaches appropriate for each; and be familiar with the editing
> approaches appropriate for different types of writing and
> documents.

> **Outcome:** Students will understand that clear communication of
> information depends on evoking the contexts in which documents
> are used and on working with readers to understand their needs.

Introduction

Cynthia L. Selfe

GOALS AND KEY FEATURES OF THIS RESOURCE BOOK

Designed for teachers of introductory technical communication courses, this book has four primary goals, each of which are tied to a key feature of the text:

- **to provide teachers key performance outcomes around which they can structure technical communication courses**: each chapter focuses on one or more key performance outcomes considered important by recognized scholars and experienced teachers of technical communication. (See the "Outcome" identified at the beginning of each chapter.) Teachers are encouraged to choose from this longer list to assemble a set of outcomes appropriate for their local institution, situation, students, and curricula.

- **to give teachers evidence they can use for shaping their technical communication class**: the resource book identifies arguments that business and industry leaders have made in support of each chapter's performance outcome. (See the section entitled "What Business and Industry Tell Us" in each chapter.) Each chapter also examines these arguments in the context of contemporary academic research in technical communication. (See the section entitled "What Academic Research Tells Us" in each chapter.)

- **to provide teachers with overviews of assignment sequences designed to address targeted outcomes with students**: each chapter of this resource book provides teachers with an overview of an assignment sequence that will help them address the targeted outcome that serves as the focus of the chapter. (See the "Assignment Sequence" section of each chapter.) This overview explains how experienced teachers of technical communication approach and work through the assignment sequence.

1

- **to provide teachers assignment materials and sample texts for each performance outcome**: in the appendices of each chapter, this resource book provides assignment sheets, activity worksheets, additional materials, and samples of student work that teachers can use in connection with the assignment. (See the appendices for each chapter.)

WHY WE CREATED THIS RESOURCE BOOK

The authors of the following chapters have written this book because, like most teachers, we care about good teaching. We also think that students in technical communication courses deserve the very best instruction our profession can provide. For this reason, we have attempted to create a book that provides teachers of introductory general education courses in technical communication with a rich series of creative and carefully crafted resources for their classrooms. We have tried to provide resources that will meet the needs of both teachers and students, and that will fill some important gaps in contemporary courses of this kind.

Why did we tackle this specific project? Why shape it as we did? The teacher/ scholars in this collection have experiences teaching technical communication in four-year- and two-year-college settings, institutions characterized by technical and liberal arts curricula, and departments that specialize in technical communication and teach that course as one among many writing courses. In doing this work, we have observed some things that have encouraged us to formulate our approach in this book:

1. We have noticed that many teachers of the introductory course in technical communication want an alternative to designing their course around the typical chapters of a textbook: writing reports, writing memoranda, creating charts and graphs, etc. In this book, we provide such alternatives, organizing each chapter around one or more key performance outcomes for students that some twenty experienced teacher/scholars consider important to address in the introductory technical communication classroom. One key goal of this book is to offer teachers a range of innovative pedagogical approaches and resources from which to choose.

We understand, of course, that good teachers will tailor their instruction to specific student populations, department goals, and institutional missions, and that they will not subscribe to *all*, or even *most*, of the outcomes we have identified in these chapters. The goal is to provide teachers with a relatively broad range of such outcomes from which they can choose some options, designing a course that meets their needs and the needs of students. Teachers can select those outcomes around which they want to focus their technical communication class, choosing from the longer list at the end of this Introduction.

2. We have noticed that many teachers often find introductory courses in technical communication less than engaging, less than amenable to creative approaches. We believe this situation persists, in part, because it is difficult to find a collection that samples some of the most engaging assignments that teachers are now providing students in these courses. Although some of this work takes place in contemporary textbooks, most of which are packed full of valuable and useful information, such

volumes are generally written by a single author and, thus, are focused around the vision and pedagogical approaches of one person, even when the author's thinking is informed by the best work in the field. While we do not want to duplicate the efforts of some very fine textbooks, we know of no resource collection that focuses on the best technical communication assignments by a range of experts in the field.

For this reason, we have designed this book to accompany a wide range of the technical communication textbooks that teachers are now using for the introductory course. We have also recruited the perspectives of specialists in technical communication, visual rhetoric, usability studies, technology studies, junior-college pedagogies, media studies, professional editing, workplace writing, collaboration and teamwork, oral communication, and writing in the disciplines.

3. We have observed that many students who take the introductory course in technical communication consider the class more as a requirement to get out of the way than a course that will prove critical to their success after graduation—and that they let their teachers know their opinion on this score! To address this situation, we have included a rich set of specific performance outcomes culled from contemporary business and industry sources. We hope that teachers can use this information to convince students of the importance of individual performance outcomes. We have also set this information within the context of academic findings that have some bearing on the target outcome. We hope such sources will provide teachers valuable help in understanding why they would want to shape a course around any particular set of outcomes. We also hope that teachers will find both the evidence from business and industry, and that from academic sources, to be useful in conversations with colleagues about the similar courses and the basic design of technical-communication courses.

4. We have noticed that some teachers of the introductory technical-communication course have few departmental and institutional venues in which they can share assignments, approaches, and questions about their courses. But good teachers are *always* interested in more good assignments and seeing what other professionals do in their courses. Teachers of technical-communication courses are no different. To this end, we have assembled assignments and instructional materials that teachers can use in the introductory technical-communication classroom to address targeted performance outcomes. We hope teachers will modify these instructional materials in ways that supplement, expand, or extend the materials contained in the textbooks they already use. And we hope they will share their successes with other teachers. We have tried to feature assignments that communicate via *multiple* channels of meaning—not only through words, but also through images, graphic elements, and design elements.

A NOTE ABOUT THE AUTHORS IN THIS COLLECTION

Although they come from institutions from around the country, all of the authors represented in this collection have taught (at one time or another and in one capacity or another) at Michigan Technological University—or they have studied with a faculty member who did.

Michigan Tech is a relatively small institution in the Upper Peninsula, but one known for its technical and engineering curricula. Its scientific and technical communication curricula has been formed, to a great extent, around the intellectual vision of faculty scholars such as Jim Kalmbach, Marilyn Cooper, Anne Wysocki, Dickie Selfe, and, Art Young, who served as the Chair of the Department of Humanities from 1978–1995. From Michigan Tech, Art went on to fill the very first endowed chair of Technical Communication in the country—the Campbell Chair at Clemson University.

Many of the contributors to this book have been graduate students in Michigan Tech's Rhetoric and Technical Communication program. Johndan Johnson-Eilola, for example, was the first Ph.D. to graduate from this program, and Stuart Selber was the second. All of these former students—whom we now are very proud to call colleagues—have made their own very productive mark on the profession of composition studies. Like Teresa Kynell Hunt and Karla Kitalong, Danielle DeVoss, Gary Bays (who was a graduate student at Michigan Tech before we even had a graduate program), Patty Ericsson, and Jerry Savage, they have authored key articles and books that have helped other teachers and scholars understand various aspects of a large and varied field. Like Michael Moore, Michael Martin, Ann Kitalong-Will, and Pete Praetorius, they have played important leadership roles of one kind or another at their institutions and in the professional organizations to which they belong. Like Jennifer Sheppard and Tracy Bridgeford, they have made new kinds of connections and contributions to our thinking and our practice. Like Summer Smith Taylor (who studied with Stuart Selber and now works with Art Young) they have become good friends and marvelous colleagues.

All of these people, we are proud to say, have helped shape the thinking of our field in highly productive ways.

We dedicate this book to Michigan Tech and to the Department of Humanities, which have given a rich set of resources to us all.

SECTION 1

Focus on Rhetorical Understanding

CHAPTER 1

Using Client-Based Writing to Teach Problem Solving

Summer Smith Taylor and Art Young

> **OUTCOME:** Students will understand the value of written communications as a problem-solving activity.

WHAT BUSINESS AND INDUSTRY TELL US

Many engineers and scientists will agree that the purpose of their work is to solve problems. Scientists search for solutions to questions about how the world works, and engineers search for design solutions that make the world work more efficiently. According to the Committee on Science, Engineering, and Public Policy (1986), "the work of managers, of scientists, of engineers . . . is largely work of making decisions and solving problems" (p. 19). ABET, the organization that accredits engineering schools, requires engineering degrees to promote "an ability to identify, formulate, and solve engineering problems" (Engineering, 2003, p. 2). The U.S. Secretary of Labor's Commission on Achieving Necessary Skills, which was intended to guide curriculum reform, listed problem solving as a foundational skill (Secretary, 1991). Other empirical studies of workplace competencies have reached the same conclusion (National Science Board, 1983; Stasz, Ramsey, Eden, Melamid, & Kaganoff, 1996). Despite their importance, however, problem-solving skills are lacking in job seekers, according to studies by organizations such as the American Society for Training and Development (Taylor, 1997) and the Society of Manufacturing Engineers (Rogers, Stratton, & King, 1999).

Writing skills are valued highly in industry. All of the sources referenced above also emphasize the importance of communication (and job seekers' lack of communication skills). The technical writing class, especially a client-based class, can serve as a site for students to learn to use writing for problem solving—individually and in teams. In fact, a survey of practicing electrical engineers found that

7

one of their main recommendations for revision of the undergraduate engineering curriculum was to "ensure that students practice cooperative problem-solving" using communication (Vest, Long, & Anderson, 1996). The engineers recommended that the students take a technical writing class taught by writing faculty, in addition to practicing writing in their engineering classes.

WHAT ACADEMIC RESEARCH TELLS US

That same recommendation is the theoretical basis of many writing-across-the-curriculum (WAC) programs as well. Proponents of "writing in the disciplines" (WID) cite the growth in rhetorical understanding and writing ability that comes from studying under a professional with discipline-specific subject-matter knowledge immersed in a particular discourse community and its established conventions. Thus, WID advocates recommend that upper-level courses in the discipline, such as "Thermodynamics and Heat Transfer" taught by a professor of mechanical engineering, be structured as writing-intensive. In such a course, writing is integral to both the processes of learning thermodynamics and to learning to write effectively about thermodynamics. Proponents of technical communication courses for engineering and science students recommend a required junior-level "Introduction to Technical Writing" course taught by that field's professionals, a course often housed in the English department. They cite the growth in rhetorical understanding and writing ability that occurs when principles of purpose, audience, genre, and media are referenced explicitly as a tool for learning, composing, and problem solving and for transferring such principles for use in different contexts, beyond specialized discourses such as mechanical engineering. An integrated WAC program, such as the one we strive for at Clemson University, embraces *both* WID and technical writing courses in order to equip students with problem-solving strategies for various and often unpredictable site-specific contexts in college and in the world of work. Our discussion in this chapter focuses on a client-based approach to rhetorical problem solving in a technical writing course composed of students from several scientific and technical disciplines and taught by a rhetorician. Because of the authentic tasks and clients to be served, our course inhabits a place between traditional, self-contained college classes and "real world" projects and audiences.

Linda Flower's empirical research, insightful theorizing, and suggestions for teaching are central to our understanding of writing as a problem-solving activity and writing as a tool for problem solving (1977, 1979, 1988, 1989, 1994). Through the examination of writers' behaviors and think-aloud protocols in workplaces, classrooms, community centers, and controlled research experiments, Flower and her colleague, psychologist John R. Hayes (1981), have described the recursive nature of the writing processes of expert and novice writers. In doing so, they have made important contributions to our understanding of how professionals-in-training, such as our technical writing students, develop into proficient problem solvers and professionals. For example, in "Rhetorical Problem-Solving: Cognition and Professional Writing" (1989), Flower deals explicitly with (1) what professionals need to know and how to do it, and (2) what and how teachers should teach so that rhetorical problem-solving strategies taught in college transfer to professional settings beyond college. In addition to recommending "realistic tasks, not toy problems" (p. 32), Flower suggests we approach the teaching of professional writing in two ways. First, "teach rhetorical strategies—both general ones and those which underlie the conventions of different kinds of discourse" (p. 33). And second, "help students develop a meta-awareness of their own strategic process," because "the expert writer is also expert about his or her own knowledge" (p. 34). Thus, involvement with course management and attention to reflective analysis aids the student's self-awareness as a rhetorical problem solver, and such meta-awareness allows writers to call upon their rhetorical knowledge when confronted with new or difficult tasks.

In addition to Flower's research, our client-based approach reflects ideas advanced in *Writing in the Real World: Making the Transition from School to Work* (1999), Anne Beaufort's ethnography of the writing and problem-solving behaviors of four mid-level employees in a nonprofit job-training agency. Beaufort's recommendations for teachers, based on her research, are similar to Flower's, yet she categorizes into five overlapping but distinct context-specific domains the metaknowledge that would be helpful to writers faced with new social contexts and writing tasks: discourse-community knowledge, subject-matter knowledge, genre knowledge, rhetorical knowledge, and writing-process knowledge. And Beaufort too recommends that we "cultivate metacognitive thinking or mindfulness, to increase opportunities for high-road transfer from one writing situation to the next" (p. 195).

We have found that designing and teaching a client-based course enables us to implement effectively the pedagogical recommendations of Flower and Beaufort as well as to respond to critical career competencies expressed by practicing engineers through ABET and the Society of Manufacturing Engineers, among others.

ASSIGNMENT SEQUENCE: FOCUSING ON PROBLEM SOLVING

Introduction

In client-based courses, students work in teams to produce print documents and electronic publications needed by clients, which may be local businesses, nonprofit organizations, public schools, or even departments in a university. Thus, a client-based technical writing curriculum provides opportunities for authentic problem solving. The client requests a document or set of documents that would help solve a pressing problem and intends to print and distribute the students' documents (or publish them on the web). For example, recent client-based projects at Clemson University have included

- brochures for parents, explaining the functions and benefits of environmentally friendly construction at a new local elementary school;

- procedures to be used for new employee training and to demonstrate compliance with the Sarbanes-Oxley act, describing the work processes of employees in various divisions of a major engineering corporation;
- proposals for University Dining Services on ways to reduce waste and increase recycling in the dining halls.

As they work on client-based projects, students are confronted with the complexities of actual audiences, actual constraints, and actual pressure to meet workplace standards. In other words, they encounter plenty of problems to solve.
We will outline a curriculum that technical writing faculty can use to help students learn to solve problems through writing in client-based classes. In brief, the curriculum involves a sequence of assignments progressing from planning and preliminary research about a communications problem, to a proposal for a solution, a deliverable that is presented to the client, and crucial metacognitive thinking throughout. This curriculum creates three interlocking problem-solving experiences. First, students practice using writing to understand the client's rhetorical context and develop a solution. Second, they learn to use writing to solve a problem for a client through the deliverable. And third, classroom management practices and other assignments and activities teach students to use communication to solve problems collaboratively within work groups and to reflect on the learning that is taking place and how the knowledge gained might be made available to them in solving future rhetorical and technical problems. **We will refer to one client-based class as an example: the class that worked with a local elementary school on brochures highlighting the environmentally friendly features of the school's construction and landscaping.** This class, like all technical writing classes in our program, enrolled 24 juniors and seniors from technical majors such as engineering, science, and agriculture.

Sequencing and Organizing Instruction

The sequence of assignments in the client-based curriculum is designed to guide students through the problem-solving process, using research and writing. The focus of the assignments that lead up to the deliverable is on understanding a problem through writing about it, rather than on implementing a solution. In the process, the assignments allow writing faculty to teach common genres and principles of composing.
The first assignment in the sequence is the preliminary research report. (See Appendix A for a sample assignment sheet that is tailored to the client-based project for a local elementary school.) The students investigate the subject matter, the needs of the audience, the expectations of the client, the perceived purpose of the document, and the nature of the genre of the requested deliverables. They typically collect information through interviews with the client, e-mail requests to experts, surveys of intended audience members, and web searches. In the case of the elementary school project, this assignment led students to interview their client, the

principal of the elementary school, school teachers, parents of children who would attend the school, and national experts on geothermal HVAC and other technical advances that are incorporated into the school's construction.

If each of the teams in the class is writing a different type of document, each team should complete its own research report; but if all teams are writing part of a larger document or part of a set of very similar documents, the parts of the report can be divided among teams and shared with the class in presentations. For example, the elementary school client requested a set of brochures describing five environmentally friendly features of the new school's construction and outdoor landscape. In this case, the class was divided into five teams (one per brochure), and each team researched a different set of questions for the preliminary research report and presentation. **The students' audience for this first assignment, then, becomes their classmates who need the information in order to complete their part of the task.** The preliminary research report unit may occupy three or four weeks of a semester, beginning after the initial weeks spent on foundational writing principles, an introduction to the project, team formation, instruction on collaboration, and so forth. In addition to principles of rhetoric, audience, subject-matter research, discourse communities, and genres such as report writing, the preliminary research report assignment provides the opportunity to teach students how to recognize a problem and to elicit information about its causes and consequences.

When beginning their research, students often approach the client expecting to be given a straightforward description of the deliverable and the information that they will need in order to complete it. Before meeting with her class's client, one student said, "We could get this all done in a week or two, I'm sure. I don't know why we need to spend the whole semester on this." Perhaps conditioned by rule-based instruction and habits of memorization that lead to success in other classes, the students tend to assume that the client has all of the answers and simply needs to be prompted to divulge them. As they conduct the research for the assignment, students are usually surprised by the ill-defined nature of the problem that they must solve. After interviewing the client, the same student quoted above expressed frustration (and demonstrated growth in her understanding of the problem-solving activity), saying "I'm not sure exactly what they want and I'm worried that what we do this semester will look unfinished because there's so much to figure out." The students are encountering, often for the first time, an ill-defined problem like those they will face on the job.

For example, one team in the class working with the local elementary school conducted their preliminary research on the client's expectations for the content and design of the brochures. The team discovered that although the client had asked the class to produce five brochures, each with an apparently different topic, the client actually saw overlap in the content that would be included in the brochures. For example, several of the brochures could reasonably cover the school's worm composting program for cafeteria waste. This team presented to the class two options for handling the overlap: first, the client could be asked to separate the content more clearly; or second, the teams with overlapping content could work together on researching and writing the parts of their brochures that overlap. After considering the preliminary research presented by a different team on the differing audiences of the overlapping brochures, the class decided on the second option.

Interviewing both the client and members of the intended audience teaches another important lesson: stakeholders in a problem often have different perspectives. The students writing for the elementary school, for example, found that the client was primarily interested in highlighting the environmentally friendly nature of the school's construction, while the principal was more interested in including general information about the school as well as counterarguments to various budgetary objections to the special construction features.

The second major assignment in the sequence is the proposal. (See Appendix B for a sample assignment sheet.) The problem/solution structure of the proposal genre engages the students in explicit consideration of the connections between the problems experienced by the audience and client and the solution that they hope to provide with their deliverable. The proposal becomes an exercise in making the problem more manageable by identifying its parts. In the elementary school example, the teams focused on showing the connections between the features of their proposed brochures and the needs of the brochure's audiences and client. They had completed independent research on the audience's needs and used the proposal as an opportunity to inform the client about their findings and how the findings support their plans. In addition to persuasion, then, the proposal assignment offers the opportunity to teach students about shaping a written deliverable to meet the needs of its readers. As the students learn more about the problem and develop a solution, they are also beginning to learn that collaboration and writing can solve problems. The focus, however, remains on thinking through the causes and consequences of the problem.

Because the proposal becomes the implicit contract between the student team, the client, and the writing teacher, the students are motivated to ensure that they have specified the parameters of their problem and solution precisely. Some teams even use the proposal to persuade their clients to narrow the scope of the project, and, as long as the teams marshal evidence behind a sound argument, these alterations can be considered an acceptable part of the problem-solving process. A client said, "It was really the students' proposals that caused me to realize I had asked for too much. They had clearly thought a lot about it and said they could really only manage this part in their timeframe, so that's what I'm now asking for."

Other assignments in the sequence can include a progress report and a variety of presentations (to the class and to the client). Faculty can ask students to describe in progress reports the problems they have encountered in the project (problems with research, teamwork, contacting the client, and so forth) and how they attempted to solve them. Students can be asked explicitly to write about and schematize what they are learning in this collaborative process that will be useful to them as researchers and writers in the future. Presentations can be focused on reporting the results of the research on the audience's and client's problems.

In thinking about their client and their client's audience, students can also reflect on broader issues such as their ethical, corporate, or civic responsibilities and the impact of their rhetorical work on culture and community. One important issue for teachers, according to Beaufort (1999), is how "to create more apprenticeship-like learning communities in our classes" (p. 194). We must figure out ways "to make the communicative work of the writing project of greater importance to students than grades" (p. 194). In a technical writing course, civic deliberations create unique challenges between the educational goals of responsible citizenship and those of vocational preparation, or as James M. Dubinsky (2004) summarizes in his "Guest Editor's Introduction" to a special issue of *Technical Communication Quarterly* on Civic Engagement and Technical Communication, "these challenges involve the complex interrelation between altruism and self-interest, between a rhetorical and an instrumental approach to teaching" (p. 245). A client-based approach to teaching technical writing, by producing real rather than "toy" problems encountered by real people in the broader community, provides a unique opportunity for examining the social implications of one's work.

The Deliverable

Writing a deliverable that will solve, or at least alleviate, the client's problem is the ultimate goal of a client-based course. As the students begin creating this product, the focus shifts from understanding the problem to implementing a solution. **Producing the deliverable, and presenting it to a grateful client at the end of the semester, teaches students that writing can make a difference by solving problems.** This view of writing helps prepare students for the workplace, where they will more typically write documents intended to solve problems than documents intended to simply demonstrate their understanding of a subject area.

Students generally begin the process of writing the deliverable with relief, believing that they are "finally able to get started." Technical students, in particular, may not value the investment in time and energy that goes into careful planning, considering alternatives, rethinking goals, and constructing subgoals. But as Linda Flower and John R. Hayes have shown, experienced and proficient writers generated far more plans and goals during the problem-solving process than did novice writers (1988). Students often think that they have already completed more than enough preparation and that designing and creating the deliverable will require simply executing their plans. However, the process of implementing the solution that has been described in the proposal teaches students new lessons about problem solving. For example, **students often discover that the solution specified in their proposal must be modified when they begin to test its implementation and receive feedback from the client and intended audience.** "You have to adjust a lot of things," one student said. "You have ideas in the proposal, but they're just ideas and once you actually start writing it, you can look at it and see if it works."

As advocated by Matson (1991) and Petroski (1992), these failures should be welcomed and investigated, not hidden. Teachers should help students see the value of the setbacks and encourage them to engage in additional problem solving based on the new information. For example, several teams in the class working with the elementary school were overly optimistic in their proposals about the amount of information they could fit into a brochure or the amount of information that would be available on a specialized topic. **Their setbacks provided opportunities for their teacher to conduct mini-lessons on research and design strategies and the use of document-design software to accommodate lesser or greater amounts of information.** At the end of the semester, all five teams presented completed brochures to the client, who immediately expressed a desire to have them professionally printed and made available in the lobby of the school (as well as distributed in other venues to a variety of audiences).

Class Management

Class management techniques that involve students in solving their own problems through writing reinforce the problem-solving focus of the course. **Students learn firsthand that writing makes a difference and causes change, and they also practice collaborative problem solving.** One effective method involves students responding on a discussion board to a question about a class management issue. For example, the teacher might ask, "How should we form the teams for the client-based project?" Students then respond with a short argument in favor of a position on the question. After the students have had a chance to respond and read others' responses, the class can vote on the options proposed. This method generally results in a near-consensus, and students learn in the process that their writing can affect their classmates' opinions and the conditions under which they work in the class. As a final task in this sequence, students describe three principles about team building that they might transfer to new situations.

A variation on this technique asks students to reflect on their work in the course in order to identify problems and test solutions. For example, a teacher might ask the students to write briefly about ways they could improve their teams' collaboration. A week later, the teacher could ask the students to write briefly about steps they have taken to implement their plans for improvement. And the next week, the teacher could follow up by asking whether the efforts made a difference. These reflections could be shared within teams or only with the teacher. When shared with the team, they engage students more fully in using writing to solve a problem rather than simply using it to think through a problem on their own. In Linda Flower's theoretical framework (1994), **students build strategic knowledge ("a trio of goals, strategies, and awareness" p. 205) through metacognitive activity ("thinking about one's thinking" p. 225),** *collaborative* **planning, and problem solving, especially when such reflective activity leads directly to action (p. 234).**

Assessment

One of the lessons students can learn about problem solving in a client-based class is that the criteria for evaluating a solution are often flexible and are always dependent on the characteristics of the problem itself. These characteristics are often viewed differently by different players in the situation, as well. The evaluation of the deliverable will generally be based on both the client's assessment and the teacher's assessment. **The client typically focuses on how well the document will solve the problem.** To receive the client's approval, the students must not only listen to the client's views about standards for the deliverable, but must also communicate information they may learn about somewhat different standards that would create a better document for the audience. In other words, **the students become responsible for negotiating some of the criteria that are used to assess their deliverables.** Establishing these criteria, or Standards for Excellence, is part of the proposal assignment (see Appendix B). Just as the students' understanding of the problem and best solution sometimes changes after the proposal is approved, the client's views may also change. As a result, the standards should be revisited frequently as the students prepare the deliverable. The students should refer to them to guide their work, revise them if necessary, and share any revisions with the client and teacher. When assessing the deliverable, the teacher may focus on how well it reflects principles taught in the course, but may also assess the extent to which the deliverable meets the Standards for Excellence established by the students and client.

The teacher and client involved in the elementary school project met frequently during the semester to discuss the students' progress. The client received and responded to the students' proposals, and the students received her feedback at the same time as the teacher's feedback and grade. The client also provided oral feedback to the teacher on the students' deliverables, and the teacher was able to take the client's perspective into account when assigning grades. The client tended to focus on the deliverables' accuracy and attractiveness. The teacher also evaluated the deliverables on course-based criteria such as use of substantive headings, establishment of a clear connection to the audience, appropriate citation of sources, and so forth. Such an arrangement is emblematic of the working lives of many engineers or scientists, in which several people work together on a proposal for funding, a proposal that must satisfy the standards of both internal managers (i.e., legal, accounting, logistical) as well as the client or funding agency.

Evaluation of the other assignments in the course, such as the preliminary research report, proposal, progress report, and presentations, need not differ significantly from ways that these assignments are evaluated in courses that are not client-based. The assignment sheets in Appendices A and B include criteria for evaluation, which can be used either to create a scoring rubric or simply to guide response. Of course faculty may select different criteria based in the emphases in their curricula.

IN CONCLUSION

This chapter describes how to use a client-based pedagogy for technical and professional writing to help students learn to approach writing as rhetorical problem solving and learn to solve the communication problems they encounter. This approach responds to industry's call for better problem-solving skills, and writing skills, in new graduates. It also draws on research indicating that reflection and metacognitive thinking help writers transfer problem-solving and rhetorical abilities to new situations, whether in other classes or the world of work. The curriculum suggested here is easily adaptable to a wide variety of client projects; faculty need not reinvent their curriculum each time they work with a new client. Although the circumstances and project requirements vary, the students always need to conduct preliminary research, write a proposal, write progress reports and reflections, and complete a deliverable. By establishing this curriculum, the teacher not only maintains consistency in his or her teaching, but also creates opportunities for students to identify and solve problems. If teachers explicitly adapt their syllabi to each new project, they are likely to resolve too many of the problems before the semester begins, thus weakening the students' learning experience. Because of the unique social and rhetorical contexts of each site-specific discourse community, college instruction can only do so much to prepare students for writing and problem solving on an as yet unknown worksite. However, **by locating client-based projects between the world of school and the world of work, we have important opportunities to facilitate the transfer of rhetorical and problem-solving knowledge from one context to another.**

APPENDIX A
Preliminary Research Report Assignment

Before you plan the content, organization, and design of the document you will write for Clemson Elementary School, you will want to know what standards the document needs to meet and what resources are available to help you produce the document. Interview members of the audience as well as our client representative, Dr. Gina McLellan, to gather the needed information. Present your findings in a report directed to me. Use the following questions to guide your research.

1. **Audience Analysis**

   For this section, you will need to interview both Dr. McLellan and members of the intended audience of your deliverable. We need to know who the likely readers of your document are. Why do they need the document? How motivated are they to read the document? What are the readers' attitudes toward your topic? Toward Clemson Elementary School? Toward environmental awareness?

2. **Client's Expectations**

   Your document will need to be approved by your client. Find out from Dr. McLellan what she expects the document to do for Clemson Elementary School. What image does she want the document to project? What expectations does she have about the content, organization, design, and length of the document? How much flexibility do you have to make decisions about the document? How, if at all, do the client's expectations conflict with the audience's needs?

3. **Available Resources**

   What sources can you consult in order to gather the information you will need to include in your final product? Name specific people and print or online materials and the types of questions they may be able to answer. Find documents of a similar purpose or audience that can serve as models for the document you will write. Start by asking your client for possible models. For each, explain the ways it seems similar to and different from the document you plan to write.

Criteria for Evaluation

- *Substance:* The report should provide thorough, detailed information that is relevant to decisions about your deliverable.
- *Organization:* The report should be divided into sections that include headings, introductions, and take-away points. The introduction of the report and of each major section should include forecasting.
- *Documentation:* All information in the report that was gathered from sources other than your team members should be properly documented so that the sources are clear.

APPENDIX B
Proposal Assignment

Write a formal proposal to Dr. Gina McLellan in which you convince her that you understand the problem that exists for the intended readers of the document you have been asked to write, and that the document planned by your group will alleviate it. Use your preliminary research findings to make your case in the proposal. Adapt the following structure as needed for your situation.

Purpose statement: In three or four sentences, state the problem faced by the audience, the consequences of the problem, and the solution that you plan to implement (the document you plan to write).

Problem and context: Describe the problem and its significance for the audience of the document.

Proposed solution: Explain how your document will solve the problem for the audience and how the document will help the school. Describe your plans for the content, organization, and design of the document, and justify your plans based on the audience's needs. Include *Standards for Excellence*—the standards that you understand must be met by the deliverable in order for them to be approved by the client.

Work plan: Establish a schedule and describe work that must be done in order to complete the project. Identify research questions—about the topic, audience, genre and/or other elements of the rhetorical situation—that you need to answer in order to produce the document. Explain how you will find answers.

Criteria for Evaluation

- *Logic:* Is there a clear relationship between the problem and the proposed solution?
- *Solution:* Will the planned document, as described in the proposal, meet the needs of the audience? Is the document thoroughly described so that Dr. McLellan can evaluate the plans and give feedback?

- *Planning:* Do you have sufficient knowledge of the background of the problem? Do you have a well-thought-out, detailed plan of action for completing the necessary work on time?
- *Persuasion:* Is each section designed to achieve a clearly articulated persuasive goal? Are claims supported by evidence? Do you use appropriate strategies to accommodate and persuade your audience?

ACKNOWLEDGMENTS

The Client-Based Program at Clemson University has been supported by funding from the Roy and Marnie Pearce Center for Professional Communication, the Bob and Betsy Campbell Chair of Technical Communication, South Carolina Sustainable Universities Initiative, and the U.S. Department of Agriculture. Case study research on the program was supported by a research grant from Clemson's College of Arts, Architecture, and Humanities.

REFERENCES

Beaufort, A. (1999). *Writing in the real world: Making the transition from school to work.* New York: Teachers College Press.

Committee on Science, Engineering, and Public Policy. (1986). *Research briefings.* Washington, DC: National Academies Press.

Dubinsky, J. M. (2004). Guest editor's introduction. *Technical Communication Quarterly, 13,* 245-249.

Engineering Accreditation Commission. (2003). *Criteria for accrediting engineering programs: 2004-5.* Baltimore, MD: Accreditation Board for Engineering and Technology, Inc. (ABET).

Flower, L. (1977). Problem-solving strategies and the writing process. *College English, 39,* 449-461.

Flower, L. (1979). Writer-based prose: A cognitive basis for problems in writing. *College English, 41,* 19-37.

Flower, L. (1988). The construction of purpose in writing and reading. *College English, 50,* 528-550.

Flower, L. (1989). Rhetorical problem-solving: Cognition and professional writing. In M. Kogen (Ed.), *Writing in the business professions* (pp. 3-36). Urbana: National Council of Teachers of English and Association of Business Communication.

Flower, L. (1994). *The construction of negotiated meaning: A social cognitive theory of writing.* Carbondale, IL: Southern Illinois University Press.

Flower, L., & Hayes, J. R. (1981). A cognitive process theory of writing. *College Composition and Communication, 32,* 365-387.

Matson, J. V. (1991). *The art of innovation: Using intelligent fast failure.* State College, PA: Pennsylvania State University.

National Science Board Commission on Precollege Education in Mathematics, Science, and Technology. (1983). *Educating Americans for the 21st century.* Arlington, VA: National Science Foundation.

Petroski, H. (1992). *To engineer is human.* New York: Vintage Books.

Rogers, D., Stratton, M., & King, R. (1999). *Manufacturing education plan: 1999 critical competency gaps.* Dearborn, MI: Society of Manufacturing Engineers.

Secretary's Commission on Achieving Necessary Skills. (1991). *What work requires of schools: A SCANS report for America 2000.* Washington, DC: U.S. Department of Labor.

Stasz, C., Ramsey, K., Eden, R., Melamid, E., & Kaganoff, T. (1996). *Workplace skills in practice: Case studies of technical work.* Berkeley and Santa Monica, CA: National Center for Research in Vocational Education.

Taylor, C. M. (1997). Back to the basics. *Quality in manufacturing.* Retrieved August 2, 2004 from http://www.manufacturingcenter.com/qm/archives/0697/697basic.html.

Vest, D., Long, M., & Anderson, T. (1996). Electrical engineers' perceptions of communication training and their recommendations for curricular change: Results of a national survey. *IEEE Transactions on Professional Communication, 39*(1), 38-42.

CHAPTER 2

Technical Reports as Rhetorical Practice

James Kalmbach

> **OUTCOME:** Students will conceptualize reports as collections of rhetorical practices and improvisational strategies. Given a specific report template or format, they will construct a document that respects that template while also respecting their material and their rhetorical purposes.

Does the following scenario sound familiar?

You are a graduate student assigned to teach a technical-writing course at the last minute. Or perhaps you are a young faculty member in a college where you teach four or five courses each semester, including a literature course, a creative-writing course, and a seemingly endless number of sections of first-year writing. And now they want you to teach technical writing. You have never worked in industry. You went right from undergraduate studies to graduate school to your present position. Still, teaching writing is teaching writing. Isn't it?

Luckily the department has already ordered a textbook for the course, and as you skim the book, the advice about report writing makes complete sense. Write a summary. Front load your points. Include background, methods, results, conclusions, and recommendations. You find good advice about graphics and page layout, and even help on creating web sites, something you always wanted to learn. This is great. Now all you have to do is figure out an assignment.

You assign the chapter on report writing that includes a clear, lucid example of a report, and you pick an assignment from the exercises at the end of the chapter.

Unfortunately, when the reports come in, they all look like the textbook's example report. They look like it, but on closer inspection, almost none of the students' topics really seem right for the pattern outlined in the book. Many parts of their documents seem to replicate other parts for no obvious reason. One student simply copied his abstract and used it again as his introduction. That can't be right, can it?

These students (like hardworking students everywhere) have taken the formula you gave them and plugged in their information in a way they hoped would please you. You have opened the bank, and they have made a deposit. The fact that their information didn't fit the book's recommended report structure is your problem, not theirs.

Now what do you do?

The result of experiences like the one described above is too often that both students and teachers end the semester disliking technical writing intensely. The students perceive the writing to be rote and mechanical: *Find the form. Fill in the blanks. Plug, chuck, and forget.* They end up feeling that writing is just another hoop to jump through, instead of seeing writing as part of the learning, knowledge, skills, and strategies that will make them productive members of their prospective profession.

And for the teacher, the experience is equally unpleasant. The classes seem rote and mind numbing: assign the chapter; grade the document. Not a hint of creativity to it. No problem solving; no new ideas; no clarity; no grace. Just toss out the guidelines and try to survive page after page of dull, uninspired prose.

To quote The Firesign Theatre, "We are all bozos on this bus."

While it is easy to recognize that such a scenario is undesirable, that **no course can prosper when neither the students nor the teacher see any real value in the course's core activities,** it is much more difficult to tease out the sources of the problem and to figure out what to do about them.

In this chapter, I argue that the central problem in the above scenario is a confusion of document templates with the rhetorical practices out of which those templates are constructed, which results in the separation of report writing from the social milieu out of which reports emerge in the everyday world of situated practice. I then present a sequence of activities that use ethnographic methodologies to bring the social context of report writing back into the classroom.

WHAT BUSINESS AND INDUSTRY TELL US

The conflict between original content and template material in report writing appears in many different forms when industry leaders talk about the importance of communication in the workplace. It appears in the criteria of accrediting agencies, in the ways leaders talk about workplace writing and learning, and in the ways companies construct style guides.

In some instances industry leaders have clearly highlighted this issue. For example, the Engineering Accreditation Commission (which accredits Engineering programs) stresses the balance between content and effectiveness in their 2005–6 general criteria for advanced level programs, which states that students should complete

. . . an engineering project or research activity resulting in a report that demonstrates both mastery of the subject matter and a high level of communication skills.

("Criteria for Accrediting Engineering Programs," p. 4)

Other business leaders emphasize these skills more indirectly by talking about real-world writing, as in this comment by a Verizon official about a vocational school technical writing program: "This course is exemplary in that it focuses on real skills for real work situations" (Kemmery & Cook, 2002, p. 32).

The tension between content and templates is also reflected in the nature of positions communication specialists get in industry. In a roundtable discussion about the importance of writing in corporate settings, Sarah James reports that one of her early corporate positions was "advising people on how to fit their scientific arguments into required formats" (Terrell, n.d.). It can also be seen in the growing importance of workplace learning programs, reflecting an awareness that much of what people learn on the job happens indirectly through water cooler conversations (Bassi, Cheney, & Lewis, 1998).

Other companies use style guides that specify formats and structures for reports as a way of attempting to teach employees how to communicate effectively to a variety of corporate audiences. These guidelines will often specify at great length the physical appearance and routing information that documents require, but provide much less information about rhetorical structure. In any large organization, such guidelines are critical to insure that documents move through the system and are acted on promptly. Boiarsky and Soven's (1995) careful case study of the documents surrounding the Three Mile Island accident demonstrates clearly that the problems with these documents were not simply matters of writing style. Rather the difficulties of the prose reflected a more complicated failure by the report writer to analyze the report writing context and to correctly route the document through that context. The Three Mile Island memo failed as much because of who it was sent to as how it was written (p. 369).

WHAT ACADEMIC RESEARCH TELLS US

Teachers of technical writing have similarly struggled with the issue of the context of writing. The initial problem that confronted my first-time teacher of technical writing was constructing a context in which the course's writing assignments were to take place. Having no experience writing in industry, this teacher relied on the textbook to create that context, but the textbook presented only a report form and a rationale for that form, not a context. The teacher quickly found herself plunged into the great paradox of the technical-writing-service course. While such courses may be designed with the goal of preparing students in majors other than English to write effectively and deliberately in professional settings, the course itself is set in a traditional classroom and is often taught by a teacher who knows little about the professions students aspire to join. Thus the paradox: **how can a technical-writing course that is deeply constrained by the contexts of schooling prepare students to write within the contexts they will encounter outside of schooling?**

This conflict between writing in school settings vs. writing in professional settings is not, of course, new. As Robert Connors (1982) has shown us, it dates back to the origins of technical-writing-services courses in engineering schools at the turn of the century. Engineering faculty wanted their students to write like engineers, while the junior English faculty members assigned to the course wanted to teach literature. The results of these conflicting value systems were not pretty.

More recently (and with much less antagonism), J.C. Mathes and Dwight Stevenson have written about this issue in their influential textbook, *Designing Technical Reports,* first published in 1976.

> In college, students write for an audience of one person—a professor; in industry, they must learn to write for a large, diverse audience in an organization. In college they write for a reader who knows the field and probably knows more about their technical material than the writer does; in industry, they must learn to write to people who perhaps do not know the field or who almost certainly know less about the material than the writer. In college, they write for pedagogical purposes—to demonstrate to a professor their mastery of concepts, processes, and information; in industry, their mastery is assumed, and they must learn to write for instrumental purposes—to help people in an organization make judgments and act upon the results they present. In short, even though engineering students may have written quite a bit by the time of graduation, little in their college writing experiences prepares them for the communication situation they face as professionals (p. xv).

In short, technical-writing teachers want students to learn how to write effectively within specific organizational contexts by taking the accepted forms of that organization and adapting them to their own content and purposes. You would like students to write the report instead of the report template writing them, but figuring out how to do so within the context of the classroom has been a vexing problem.

Over the years many different approaches to resolving the paradox of context have been tried, abandoned, and tried again. One solution is to limit a section of a general technical-writing-service course to students from a specific major. This move enables the teacher of that course to focus more narrowly on the professional concerns of that discipline.[1]

Another strategy is to use case studies: teachers teach genres such as report writing or issues such as ethics by having the students work through and write about a fictional professional setting.[2] A popular variation on this approach is to build cases out of the real documents that surround major disasters. These documents are included in the appendices of the national reports issued by commissions charged

---

[1] See for example, Kumpf and Emanuel's (1996) description of a technical-writing course designed for Industrial Engineering students.

[2] Case studies often are used to examine issues such as international technical communication or ethics that might not ordinarily emerge in the normal course of technical writing projects (see Bosley, 2000; Frantz, 1999; Kynell & Stone, 1999).

with identifying the causes of the disaster and recommending changes to avoid such a disaster in the future. Using real documents has the advantage of situating discussions of writing within the context of real-world outcomes.[3] Savage and Sullivan (2001) published a popular variation on the case method by collecting stories from professional technical writers that depict various aspects of their lives at work. These stories provide a way of bringing the experiences of working technical writers into the classroom in order to contextualize student projects.

Other approaches to the paradox of context include focusing the course on media projects such as brochures, booklets, flyers, videos, and web sites (see Addison, 1982; Bernhardt, 1986; Kalmbach 1988; Kramer & Bernhardt, 1999) as well as service-learning approaches that have students work with real clients outside the classroom (see Bowdon & Blake, 2003; Stone, 2000; Wickliff, 1997).

TECHNICAL WRITING AS SITUATED LEARNING

Recent work on teaching the technical-writing-service course has used activity theory as its theoretical lens.[4] Lave and Wenger (1991), for example, examined the concept of "situated learning," which they argue is

> . . . the basis of claims about the relational character of knowledge and learning, about the negotiated character of meaning, and about the concerned (engaged, dilemma-driven) nature of learning activity for the people involved. That perspective meant that there is no activity that is not situated. It implied emphasis on comprehensive understanding involving the whole person rather than "receiving" a body of factual knowledge about the world; on activity in and with the world; and on the view that agent, activity, and the world mutually constitute each other (p. 33).

They go on to argue that learning, outside of the classroom, takes place in a community of practice.

> As an aspect of social practice, learning involves the whole person; it implies not only a relation to specific activities, but a relation to social communities—it implies becoming a full participant, a member, a kind of person. In this view, learning only partly—and often incidentally—implies becoming able to be involved in new activities, to perform new tasks, and functions, to master new understandings. Activities, tasks, functions, and understanding do not exist in isolation; they are part of broader systems of relations in which they have meaning. These systems of relations arise out of and are reproduced and developed within social communities, which are in part systems of relations among communities (p. 53).

---

[3] Boiarsky and Soven (1995) end their textbook with case studies of the Three Mile Island nuclear accident, the Challenger disaster, and the African slave burial ground dispute. The Challenger chapter includes reproductions of 12 separate reports and documents from the disaster.

[4] See Lave and Chaiklin (1993), Lave and Wenger (1991), Orr (1986, 1996), Schon (1983), and Suchman (1987).

For technical-writing-service courses, activity theory argues that **to write a report is to make visible the complex social practices of everyday organizational lives. Effective reports can best be understood within the context of those social practices.** This view of learning as socially situated action has translated into technical-writing-service courses that insist students find topics for technical reports that are located in real social situations.[5]

ASSIGNMENT SEQUENCE:
TEACHING REPORT WRITING AS SITUATED ACTION

Viewing reports as situated actions is the approach to teaching that I have taken in my technical-writing-service courses. I am strident in insisting that students locate report topics in concrete and real social contexts outside of the classroom (see Appendix A), and I help them find alternatives when it is clear that they genuinely can't situate their work on their own.

The initial step in this report-writing unit is sustained and intensive brainstorming. I ask students to list 25 organizations they have been involved with. These organizations can be school related or work related. The group could be here or at home. It could be a social group, an extracurricular group, or a religious group. I ask that they make up their list quickly and uncritically, taking no more than 10 minutes, and to really try to come up with 25 organizations.

Students bring these lists to class and then narrow them down to the five or so organizations that they would really like to work with. The students then write about problems or issues that these groups are encountering, comparing those issues to the criteria in Appendix A to determine whether an issue has the potential to be a report topic (always keeping in mind that many if not most of the world's problems cannot be solved by writing a report). I try to get students to use this process to come up with three to five possible topics because the act of comparing possible topics helps them to make better choices.

The next step has consistently been the most valuable part of this assignment. **Once students have chosen a potential report topic for a real organization, they must interview (ideally face to face) the person who would make a decision on whether or not to act based on the proposed report.** (For a description of this interview process, see Appendix B.) Talking face to face with a real person seems to help students situate their projects in social practice better than any other form of analysis that I have tried.

Once students have a topic to research, a real social context in which to situate their report, and a decision maker to whom the report should be directed, I ask them to articulate a rhetorical purpose for including each of the components of a report format. Starting with the summary, students must construct—for their topic, their research, and their decision maker—a purpose for writing the summary: a specific goal they wish to achieve as well as the reasons

---

[5] See Artemeva, Logie, and St-Martin (1999), David and Kienzler (1999), Freedman and Adam (1996), Kain and Wardle (2005), and Wahlstrom (2002).

they believe the decision maker would read and use the summary. We complete this process for each of the standard elements of the report: the background, methods, results, conclusions, recommendations, and appendices. When we are done, students have a situational analysis from which they can produce an initial draft of their report and also explain why are they are choosing to include these particular report elements.

The goals of these prewriting activities is to put students in a position to write a report that honors their material rather than forcing that material into an inappropriate format. **When students have a clear rhetorical purpose for why they are writing, they are in a better position to negotiate between the requirements of their organization and the needs of their content.**

Appendix C includes a student report that is an example of the end result of this process. Recall that students are asked to situate report topics within the social contexts and practices of their own lives. This student chose to write about her church. She was a Sunday-school teacher who felt that the church was not preparing her well enough for the classroom. She surveyed her fellow teachers and found that many agreed with her. She then called lead teachers from other churches in the area and found a Sunday-school teacher development model that she felt would be likely to succeed at her church.

The report that she wrote uses an inductive narrative structure. She starts by describing the personal experience that inspired her to do the research, she presents the results of her survey of other Sunday-school teachers, and then her research as to what other churches are doing in terms of professional development. The report moves in a spiral, starting with her experience and moving to wider and wider contexts, ending with her recommendation. It is a report structure she arrived at inductively by relating her purposes to her content and her audience. **You would be unlikely to find such a report structure described in a technical-writing textbook, but it was extremely effective for this particular project.**

THE IMPORTANCE OF REMAINING FLEXIBLE

It is not unusual for experienced teachers of technical-writing-service courses to have tried all of the approaches to bringing professional contexts into their courses that have been described in this chapter over the course of their career. Often we end up with courses that cobble together different aspects of these approaches in interesting (and productive) ways. What we too often lack is a theoretical lens to help predict and guide our explorations. **When we don't have a theoretical lens, we don't have a basis from which to troubleshoot when things go wrong and to improvise when students (through the process of planning and writing their projects) take a course in an unexpected direction. Without theory, we struggle to learn from our mistakes.** I have found activity theory to be a particularly fruitful way of understanding report writing in real social contexts. But the important step isn't to agree with me, it is to articulate your theoretical foundations so you can build on them.

Whatever theoretical approach you take, whatever goals you articulate, whatever values you embrace, the most effective technical-writing teachers (like the most effective technical writers) are both theoretically informed and flexible in practice, ready to switch approaches in response to the emerging complexity of the classroom. Teaching technical writing should not be the rote assignment of forms and the production of dull lifeless prose poured into those forms (as it was in the story that began this chapter). Such approaches result neither in interesting classes nor students prepared to write outside the classroom. Instead, **teaching technical writing should be an act of continual creativity,** just as effective writing within real social contexts is an act of continual creativity, weighing the needs or organizations, templates, and decision makers against the goals of writers and constantly changing approaches to keep those needs in land goals.

APPENDIX A
Report Writing as Situated Practice Assignment

The Report Project

Your task for the report project involves writing a report geared to a specific organization that focuses on an issue that is within the domain of that organization to solve or act on. It should be a specific, narrowly focused problem; one that you can research within a three week unit so that you will be able to write a document that makes recommendations based on your research.

The best reports are for organizations you care about or would like to see get better. Ideally you should know the people in the organization, but you can also act as a consultant working for an organization you do not belong to, and interview members to get the information you need for the planning memo.

Here are some guidelines for picking report topics:

Promising Report Topics	Problematic Report Topics
Have a clear organizational context and a narrow, specific purpose.	Involves an issue that concerns the writer but has no clear organizational context.
Involves an issue that is within the domain of the organization to act on and solve.	Focuses on issues that cross many different organizations and cannot be solved by any one of them.
The issue is specific, narrowly focused, and concrete.	Involves large or philosophical issues.
If money issues are involved, they can be reduced to a concrete budget that is within the scope of the organization to justify and underwrite. Example: Which printer should our student group buy?	Often involves money issues of unknown costs or may require approval of unknown groups beyond the scope of the current organization. Example: Our unit needs to pay employees more.
Involves procedures, equipment, or specific questions that can be easily researched.	Involves people issues without a clear focus (morale is low: why my supervisor shouldn't treat me so poorly).

While these guidelines may be foreboding, many projects can be adapted to them. If you have a clear issue and a clear organization, but all of your solutions seem beyond the limitations sketched above, then think about how you can focus on a smaller chunk of the bigger problem. Find research that has a narrow focus but can contribute to a solution further down the line.

APPENDIX B
Bringing Social Context into the Report Assignment

INTERVIEW WITH A DECISION-MAKER

Your first project in the report unit is to interview a decision maker about a potential report project. The decision maker is the person (or persons) who would make a decision on whether to act on the recommendations in your report. The purpose of the interviews is to provide you with context about the problem, the consequences, and the nature of the evidence that would likely make a difference in his or her decision-making process.

Conducting Decision-Maker Interviews

If several different people would be making a joint decision on whether to continue, you only need to interview one of those people; however, you can also do a group interview. For example, if you are writing a report for a fraternity or sorority, you may find it more useful to get the executive board together for a group interview rather than just interviewing the president. You can do your interview face-to-face, over the phone, or over the Internet, but f2f is preferred. You will consistently get the best, most reliable information from f2f interviews. Here is the critical information that you will need to gather during the interview:

1. In the decision maker's view, what is the nature of the problem? In too many reports, the report writer identifies a particular issue as a key problem for the organization *from his or her own personal perspective* rather than from the perspective of the decision maker/reader. The decision maker you interview may not agree that the issue you have identified is a problem for the organization. If you and your audience do not agree about the nature and scope of the problem, you are unlikely to agree about solutions, and your report is likely to fall on deaf ears. Start by asking your decision maker to describe the problem/issue in his or her own words.
2. What are the consequences of the problem to the organization? Consequences help determine whether the benefits of solving a problem outweigh the costs.
3. What sort of information will help the decision maker evaluate your report?
4. What criteria will the decision maker use in evaluating your report: Cost? Feasibility? Time to implement?

5. Are there any landmines? Landmines are unforeseen issues that lurk beneath the surface of a project suddenly exploding in your face when stepped on.
6. If you do not know the decision maker you are interviewing, ask them to describe their background and whether any aspect of that background is likely to cause them to say no to your project.

Interviews Tips

- Start with your most open-ended questions and end with your most specific questions. Don't, however, be a slave to your questions. If an interesting issue comes up that you did not anticipate, follow up. If you do not get to all of the questions, do not worry about it.
- Try to find a quiet place to conduct your interview, one that will minimize interruptions. If you are in a public room, sit in a corner with your back to the door.
- Use silence. When your interviewee stops, don't rush to fill the void. Wait a bit to see if he or she volunteers anything else. If silence doesn't work, ask a follow up question, or make a general request such as "Is there anything else you can add?"
- Avoid answering questions. If you answer questions, you become the expert and you lose control of the interview. Put the subject off until the end of the interview. If your subject gets you answering the questions, you are in trouble.
- You may take notes during the interview or immediately after the interview, write down everything you can remember. Write down as many direct quotes as possible.

Writing up the Interview

After you have conducted your interview, write it up and turn it in to me. Here are some notes on what to do and what I need to see.

- Write your report as a single-spaced memo.
- Start with a background section about the project.
- Include a section on the details of the interview. I need to know the specifics: when the interview took place, where it was held, how long it went, and in general how it went.
- Include a section on results, describing what you learned. I do not want a transcript of the interview. Instead, summarize the interview and include a few direct quotes. Direct quotations are the most effective way to present interview data; such quotes make the subject come alive for your reader.
- After presenting the interview, step back and think about what you learned and what plans you have for your report as a result of these interviews.
- Attach your schedule of interview questions to the memo as an appendix.

APPENDIX C
Sample Student Paper

March 19, 2002

To: Jane Smith
        Normal Church of Christ Sunday School Coordinator

From: Joyce Wilson
        Primary Sunday School Teacher

Re: Teacher Development at the Normal Church of Christ

Summary

The Normal Church of Christ Primary Sunday school teachers are volunteers, many who have had no formal teaching experience. Although a year-long Teacher Development class is available, most teachers are unable to attend this class before they begin teaching. In addition, the primary school classes and the Teacher Development class meet at the same time, thus the teachers cannot take the class while they are teaching primary children. I have completed a study of what teacher development needs the Normal Church of Christ primary teachers would like addressed. I also researched how other churches in Central Illinois prepare Primary teachers for the classroom experience. The purpose of this report is to present the results of this study and recommend that our Primary teacher be given a mini-class in teacher development to help teachers feel more prepared when they begin teaching. Out of the twelve teachers surveyed, only one had taken the Teacher Development class before they began teaching in Primary. Because most of our current teachers have not taken the Teacher Development class, 9 out of 11 teachers felt that they did not feel prepared for the classroom experience.

The mini-class, based on a model I found in Peoria, would focus on 1) roles of the Primary administration and the church librarian and how each of them plays a part in assisting the teacher and 2) a discussion of different ideas of how to relate to and teach the children. All eleven teachers surveyed who had not previously attended the year-long Teacher Development class indicated that they would be interested in such a mini-class.

Background

I was asked to serve as a primary Sunday school teacher in January 1992. I met with Jack Elliot on the Tuesday before I was to begin teaching. At this meeting, I was introduced to my team-teacher, who was also a new teacher and had no teaching experience. We were given a manual with the lesson plan for the year, and a handout about how to handle discipline problems. Because I had not taken the year-long Teacher Development class; I was very apprehensive about teaching for the first time. I did not have any idea of what type of behavior to expect from the children, or what teaching aids would be best to hold their attention. I discussed my apprehension with my team teacher. She told me that she also felt ill prepared to begin teaching the following Sunday. The first class, not surprisingly, did not go very well. We presented the lesson in segments that were too long for our students' attention span, and we did not use many visual aids. Our students were restless and did not seem to learn much. It has only been recently that I have stumbled upon some teaching aids that have helped the children be more responsive to the lessons. I began thinking of how helpful it would have been to have attended a mini-class for new Primary teachers before I had entered the classroom.

Because of my teaching experience, I wondered a) if I was the only teacher who felt ill-prepared to teach and b) if other teachers felt that they would have benefited from some sort of min-class on teaching before they had to teach. I decided to survey the other teachers about their feelings on teacher development. This survey was given in February 2001.

Results of Teacher Survey

Out of the 12 teachers surveyed, only one teacher had taken the year-long Teacher Development class. She had found this class very helpful. This teacher indicated that two main points that had helped her to be a better teacher was the use of visual aids, and how to relate to different age groups.

The other 11 teachers surveyed had not taken the Teacher Development class. In response to the question "Did you feel prepared for your first classroom experience?" 82% (9/11) did not feel prepared, and when questioned "As an incoming teacher, would you have been interested in a mini-class on teacher?", 100% of the teachers indicated that they would have been interested in a mini-class had one been offered.

Based on this survey, I knew that I was not the only teacher who had felt ill-prepared. I also knew that a mini-class appealed to most of the other teachers. But our church did not have one, and has not ever had one. Because of the time conflict with Primary, a mini-class for the Primary teachers would be more feasible than the Teacher Development course. I decided to talk to Primary Inservice Leaders in other churches in the region to see if they had any type of mini-class or teacher preparation for their Primary teachers before they started teaching.

Results of Phone Survey to Other Central Illinois Churches

I contacted all 10 Primary Inservice Leaders in our region during the week of February 16th through the 23rd. Nine out of the ten churches did not have any programs for incoming Primary teachers. But most of the Inservice leaders felt that some type of program would be helpful to teachers. One Inservice leader in Champaign said that she felt that if they offered some type of program to incoming teachers, their teacher turnover might decrease. Another Inservice leader in Pekin said that they did not have a program, but that if they did, more people would accept positions in Primary. This particular Inservice leader also indicated that she had heard of a mini-class on Teacher Development that the Peoria Primary Inservice Leader had started recently.

I contacted the Peoria Primary Inservice Leader and found that the Peoria did in fact offer a Teacher Development mini-class. This three hour mini-class is offered semi-annually in September and January, which is when they have the greatest number of new teachers. They invite all of the incoming teachers, and any current teachers who may have not previously had the opportunity to attend the mini-class. On the average, according to the Inservice Leader, 6 Primary teachers attended. The Primary Inservice Leader organizes and oversees this mini-class.

The first hour of the class, the teachers were taken to the gym. There are several tables set up around the room in a semi-circle, with the Primary President at one, and a table for each remaining members of the Primary Presidency: the first and second counselors, and the secretary. The Inservice Leader also has her own table. In addition, two Primary teachers sit at tables displaying different teaching ideas. The teachers spend approximately 10 minutes talking to each person in the room. The teachers are free to ask any question they may have, and the person at the table explains what their particular duties are, and what role they play in helping teachers. The Primary Presidency Secretary, for example, hands out a list of names and birthdays of each child, explains any special needs of the children that will be in the teacher's class. The teacher asks some specific questions about a particular child. The teacher then rotates to the next table, and begins the process again.

Next, there are two 1-hour classes on various teaching ideas. One class is given by a teacher and the other is given by the Primary Inservice Leader.

This Primary Inservice Leader was very enthusiastic about their program. She said that several of the teachers came up to her after a couple of months of teaching and told her how much the mini-class had helped them.

After talking with the Inservice Leader, I could see that a mini-class could be developed to meet the needs of our own church.

Recommendations

The Normal Church of Christ Primary teachers have indicated their interest in a mini-class on Teacher Development. We could incorporate ideas from our teachers with the program in Peoria. I think that we should offer a class semi-annually. We also see our greatest number of new teachers in September and January. A mini-class would begin with tables in the gym, followed by a library tour, and concluding with two classes on teaching ideas of the classroom, one taught by the inservice leader, and one taught by a current Primary teacher.

We could set up tables in the gym for each member of the church council. You may want to emphasize to the members that the teachers would like to know what expectations he/she has of the teachers, and what policies that particular person may oversee. From here, the teacher would rotate to the next table, until each teacher had visited every table.

After this session, the librarian should give teachers a tour of the church library and show them how many resources are already prepared, and available for teachers to use in their classes. I met with the librarian when I was searching for resources for my own classes, and I was amazed at how many visual aids were already prepared. By using resources that had already been prepared, I was then able to concentrate on other areas of my lesson. The library also contains flannel boards, crayons and chalk, and handouts to be copied. The year-long Teacher Development manual and other Teacher Development books are available for check-out from the library, as are several short videos for lesson variety.

After the library tour, you, as the Inservice Leader, could give a lesson on the age characteristics of children. At the library, I found a book, "The How Book for Teaching Children." At the end of the book, there is a chapter about the age characteristics of children from 18 months to 12 years, and what type of activities are most appropriate for any particular age group. This book also has many ideas on how to prepare for your first class, classroom management and how to help the children really learn what the teacher is teaching. I think that it would be an excellent book to hand out to all incoming Primary teachers.

For the final hour, you could select a teacher who has been teaching for a while to teach a class on how to get and keep the attention of the children. I would recommend, Brian Edwards. He has been teaching for about four years and most of the children can't wait until they will be old enough to attend his class.

Conclusion

Although most teachers would like the opportunity of attending the year-long teacher development class, they will not be able to take it while they are teaching Primary. A mini-class, however, is a viable option. I received positive feedback from most of the Normal 1st teachers when they were surveyed about the possibility of a mini-class. Most of the other churches that I talked with clearly felt that by having a mini-class would help a teacher feel more prepared to begin teaching in Primary. Our church can and should develop a mini-class that would help prepare incoming teachers for teaching. By offering such a mini-class, we can help the classroom become a better place for the teachers and, most importantly, for the children.

REFERENCES

Addison, J. (1982). Brochures: A teaching rhetoric. *The Technical Writing Teacher, 10,* 21-24.

Artemeva, N., Logie, S., & St-Martin, J. (1999). From page to stage: How theories of genre and situated learning help introduce engineering students to discipline-specific communication. *Technical Communication Quarterly, 8*(3), 301-316.

Bassi, L., Cheney, S., & Lewis, E. (1998). Trends in workplace learning: Supply and demand in interesting times. *Training & Development, 52*(11), 51-75.

Bernhardt, S. (1986). Seeing the text. *College Composition and Communication, 37,* 66-78.

Boiarsky, C., & Soven, M. (1995). *Writings from the workplace: Documents, models, cases.* Boston, MA: Allyn & Bacon.

Bosley, D. (2000). *Global contexts: Case studies in international technical communication.* New York: Pearson Education.

Bowdon, M., & Blake, S. (2003). *Service-learning in technical and professional communication.* New York: Longman.

Connors, R. (1982). The rise of technical writing instruction in America. *Journal of Technical Writing and Communication, 12*(3), 329-351.

Criteria for accrediting engineering programs. (2004). Baltimore, MD: Engineering Accreditation Commission.

David, C., & Kienzler, D. (1999). Towards an emancipatory pedagogy in service courses and user departments. *Technical Communication Quarterly, 8*(3), 269-283.

Frantz, A. (1999). *Cases in technical communication.* New York: Wadsworth.

Freedman, A., & Adam, C. (1996). Learning to write professionally: "Situated learning" and the transition from university to professional discourse. *Journal of Business and Technical Communication, 10*(4), 395-427.

Kalmbach, J. (1988). Technical writing teachers and the challenges of desktop publishing. *The Technical Writing Teacher, 15*, 119-131.

Kemmery, R., & Cook, H. (2002). Written communication skills for the 21st century. *Techniques, 77*(4), 32-34.

Kramer, R., & Bernhardt, S. (1999). Moving instruction to the web: Writing as multi-tasking. *Technical Communication Quarterly, 8*(3), 319-336.

Kain, D., & Wardle, E. (2005). Building context: Using activity theory to teach genre in multi-major professional communication courses. *Technical Communication Quarterly, 14*(2), 113-139.

Kumpf, E., & Emanuel, J. (1996). "Stepping lively": Reformatting the gap between student writing and professional writing. *Technical Communication, 43*(2), 164-171.

Kynell, T., & Stone, W. (1999). *Scenarios for technical communication: Critical thinking and writing.* Boston, MA: Allyn & Bacon.

Lave, J., & Chaiklin, S. (Eds.). (1993). *Understanding practice: Perspectives on activity and context.* New York: Cambridge University Press.

Lave, J., & Wenger, E. (1991). *Situated learning: Legitimate peripheral participation.* New York: Cambridge University Press.

Mathes, J. C., & Stevenson, D. (1976). *Designing technical reports: Writing for audiences in organizations.* Indianapolis, IN: Bobbs-Merrill.

Orr, J. (1986). Narratives at work: Story telling as cooperative diagnostic activity. *Proceedings of the 1986 ACM conference on Computer-Supported Cooperative Work* (pp. 62-72). New York: ACM Press.

Orr, J. (1996). Talking about machines: An ethnography of a modern job. Ithaca, NY: Cornell University Press.

Savage, G., & Sullivan, D. (2001). *Writing a professional life: Stories of technical communicators on and off the job.* Boston, MA: Allyn & Bacon.

Schon, D. (1983). *The reflective practitioner: How professionals think in action.* New York: Basic Books.

Suchman, L. (1987). *Plans and situated actions: The problem of human-machine communication.* New York: Cambridge University Press.

Stone, E. (2000). Service learning in the introductory technical writing class: Perfect match? *Journal of Technical Writing and Communication, 30*(4), 385-398.

Terrell, C. (n.d.). *Writing for corporations.* The Woodrow Wilson National Fellowship Foundation. (Online). 10-29-04. http://www.woodrow.org/phd/WRK4US/WRK4US_Discussion_Index/corporate_writing html

Wahlstrom, R. (2002). Teaching the proposal in the professional writing course. *Technical Communication, 49*(1), 81-88.

Wickliff, G. (1997). Assessing the value of client-based group projects in an introductory technical communication course. *Journal of Business and Technical Communication, 11*(2), 170-191.

CHAPTER 3

Using Multiple-Source Research to Create Persuasive Evidence-Based Communication

Jennifer Sheppard

OUTCOME: Students will understand the value of conducting both primary, experientially-based research and secondary-source research as a means for constructing rhetorically situated, persuasive communication.

INTRODUCTION

In light of the complex workplace communication demands students now face when they graduate, a key focus for contemporary technical-communication classes involves audience-centered communication that teaches students how to analyze institutional problems, to conduct multiple source research, and to use that work to create persuasive, evidence-based documents. This chapter focuses on the rhetorical practices and values of one case-study profession (i.e., nurses) to explore current research on rhetorical practices and provide readers with a sample sequence of assignments. The discussion that surrounds this case study and the pedagogical activities I have derived, however, are applicable to a broad range of disciplines and technical-communication classes.

Looking at industry research in both Business and Health Care, as well as Professional Communication pedagogy and theory, this chapter considers the ways in which students must be equipped for the written, oral, and visual communication demands they will encounter on the job. **The assignment sequence helps teach students how to conduct both primary, experiential-based research, and traditional published scholarly research, using that work to produce compelling communication. It involves students in examining workplace documents,** conducting interviews and observations, and doing traditional academic research—with the goal of teaching them to gather and use research to negotiate the rhetorical situations within which they communicate.

WHAT BUSINESS AND INDUSTRY TELL US

Professional Communication in Business and Industry

In a survey recently released by the National Commission on Writing—*Writing: A Ticket to Work . . . Or a Ticket Out: A Survey of Business Leaders*—Human Resource (HR) Directors at 120 of **America's top corporations reported that "Writing skills are fundamental in business" (2004, p. 10). Responsible for the employment of nearly eight million Americans, half of the HR Directors who responded reported that they "take writing into consideration when hiring professional employees" (p. 5).** "In most cases," said one respondent, "writing ability could be your ticket in . . . or it could be your ticket out" (p. 5). In fact, many Directors commented that poorly written application materials rarely resulted in further evaluation of job applicants. As for writing's impact on advancement, one respondent noted, "It's really not a promotion issue since you'd never get to the point of promotion without good communications skills. You can't move up without writing skills" (p. 18).

This survey illustrates the extent to which writing is a part of everyday life in the workplace, no matter what the profession. According to another survey participant,

> All employees must have writing ability. Everything is tracked. All instructions are written out. Manufacturing documentation, operating procedures, reporting problems, lab safety, waste-disposal operations—all have to be crystal clear. . . . They must be able to communicate clearly, relay information, do postings, and the like (p. 10).

In professional positions, the writing and communication practices of employees come under even more scrutiny from management. One respondent commented that "It's increasingly important to be able to convey content in a tight, logical, direct manner, particularly in a fast-paced technological environment" (p. 10). Another contended that "good writing is a sign of good thinking. Writing that is persuasive, logical, and orderly is impressive. Writing that's not careful can be a signal of unclear thinking" (p. 10).

As important as written communication skills are in a range of fields, however, employers responding to the survey reported difficulty in finding employees with adequate preparation in writing and communication tasks. According to respondents, this seemed particularly true of new college graduates. One participant commented that "People's writing skills are not where they need to be. Apart from grammar, many employees don't understand the need for an appropriate level of detail, reasoning, structure, and the like" (p. 16). Another respondent suggested that "Recent graduates may be trained in academic writing, but we find that kind of writing too verbose and wandering" (p. 16). In other words, despite employers' desires for employees that can create organized, persuasive, rhetorically situated communications, most find newly hired college graduates underprepared for these demands.

One case-study example of the demands for effective professional communication in the workplace can be seen in the field of Nursing and Health Science. Mary Chaffee (2002), Deputy Directory for the Navy Medicine Office of Homeland Security, contends that for nurses, "communicating effectively in a wide variety of environments and situations is critical to achieving professional, political, and policy objectives" (p. 93). As the primary patient advocate, they are in a unique position to observe and understand how policies and procedures affect not only patients, but also staff and the health-care institutions for which they work. However, in order for nurses to accomplish these objectives, they must develop skills for appropriately and persuasively tackling these issues. Not only does this development entail learning how and where to look for evidence, it also requires learning how to analyze problems and construct ideas in ways that will convince others to take the desired action.

Nursing Researchers Judith Leavitt, Sally Cohen, and Diana Mason (2002) argue that "Nurses engaged in the politics of the policy process will find that their efforts are most effective when they systematically analyze their issues and develop strategies for advancing their agendas" (p. 71). They go on to suggest that what "nurses bring to the deliberation of any health policy issue [is] an appreciation of how policies affect clinical care and patient well-being" (p. 71). Therefore, a primary goal for any educator preparing future nurses or health-science students for their profession should be to help them learn how to combine practical experience and observation with rhetorically based ways of analyzing and arguing. **Like students from all technical disciplines, nursing and health-science students need to learn how to conduct research, gather evidence through multiple sources, and construct persuasive arguments in collaboration with others. They need to know how to use written language to create positive changes for clients and patients and within institutions.**

WHAT ACADEMIC RESEARCH TELLS US

Professional and Technical Communication Pedagogy, Research, and Theory

Not surprisingly, these findings from business and industry and by health-care researchers are among the top pedagogical and theoretical concerns of educators in Professional Communication. For example, in the introductory chapter of the fifth edition of his popular textbook, *Technical Communication: A Reader-Centered Approach*, Paul Anderson (2002) writes the following to students:

> **From the perspective of your professional career, communication is one of the most valuable subjects you will study in college.** . . . Looking around [at work], you'll discover ways to make things work better or do them less expensively, to overcome problems that have stumped others, or to make

improvements others haven't begun to dream about yet. Yet all of your knowledge and ideas will be useless unless you communicate them to someone else (p. 4).

Anderson goes on to provide several illustrations of how valuable written and oral communication are to a diverse set of scientific and technical fields. One particularly poignant example he provides is of Larry, a new hospital dietician. Larry, he writes, "must communicate to accomplish his goal of increasing efficiency in the hospital kitchen where he works. Larry's ideas will reduce costs and improve service to patients only if he presents his recommendations persuasively to people who have the power to implement them" (p. 4).

Referring to this example, Anderson highlights three critical ways, among others, in which the writing of professionals differs from that of students. Each of these is a significant dimension of workplace communication that must be addressed in some manner by Professional/Technical Communication pedagogy. Anderson says that workplace communication

1. *serves practical purposes:* "On the job . . . readers are coworkers, customers, or other individuals who need the writer's information and ideas in order to pursue their own practical goals" (p. 5).
2. *addresses complex audiences:* "At work . . . you will need expertise at constructing one communication that simultaneously satisfies an array of individuals who will each read it with a different set of concerns and goals in mind" (p. 6).
3. *is shaped by social and political factors:* "Every communication situation has social dimensions. . . . You will need to attune the style, tone, and overall approach of each communication to these social and political considerations" (p. 8).

By helping students learn to identify and address issues of purpose, audience, and context, Anderson argues that they will be well on their way to being able to produce effective professional communication.

The pedagogical goals cited by Anderson are supported by evidence from numerous research projects and longitudinal workplace studies. Anthony Paré's 2002 article on participatory action and workplace communication provides a lucid example of just how complex the rhetorical dimensions of professional writing can be. Discussing his writing consultancy work with social workers from small, isolated communities in Northern Canada, he demonstrates how success in communicating and bringing about change are directly related to communicators' abilities to respond to the purposes, audiences, and situations in which they find themselves. He contends that **"Workplace communications are fraught with implications for relationships within institutional hierarchies, and the failure to understand those dynamics reduces credibility and effectiveness" (p. 70). Only through learning to recognize and adapt to such local and situated practices can**

one come to understand how they operate and how they shape the kinds of communication that are acceptable within a given context.

What this came to mean for Paré's pedagogy was that rather than beginning his workshops with a preconceived plan for what students would write about, he needed to structure the course in such a way that the students could first identify and then address communication problems relevant to *their* particular communities and contexts. Through follow-up surveys and interviews, he found this contextualized strategy to be far more successful in long-term implementation because it allowed participants to account for local constraints and expectations.

Writing about how this understanding of the situatedness of professional writing shaped his new pedagogical practices, Paré (2002) explains that workplace literacy

> in all of its specialized disciplinary and institutional manifestations is always local and located (never universal, never transcendent), always saturated with values and beliefs (never neutral, never merely a vehicle for communication), and always shaped by and for particular social actions and ends (p. 63).

What is of value here about Paré's work is his emphasis on understanding how **localized contexts shape the persuasiveness and success of communication within particular professional settings.** Further, it is only through identifying and addressing the unique rhetorical parameters of a workplace that a communicator can come to create texts that effectively satisfy the writer's objectives.

In another relevant workplace study, researcher Jamie MacKinnon (2003) spent 20 months following the changes in writing of 10 newly hired economists and analysts at the Bank of Canada. Through interviewing participants and their managers several times and scrutinizing the writers' work, MacKinnon was able to identify many significant changes in communication habits that occurred over this timeframe. One of the most notable transformations that came through experience was the writers' ability to understand better the "social and organizational context" (p. 414). MacKinnon notes that while many newly hired participants expressed confidence in their ability to address the diverse audiences of the bank in their first interview, most looked back on these comments with skepticism a year later (p. 415). MacKinnon writes,

> A critical aspect of the participants' development appeared to be learning a good deal about the social and organizational context in which they wrote: the business functions performed by their departments, the job performed by their readers (and thus their readers' needs for information), and the values and beliefs implicit in the organization's activities (p. 414).

In part because they gained a greater understanding of audience, purpose, and context, the participants in MacKinnon's study also became more skilled in crafting credible and persuasive documents. At the end of the study, **almost all of the writers reported that they had become better at using the feedback of coworkers and bosses to strengthen their work. Doing so helped them to more effectively**

anticipate objections and expectations of readers and, therefore, to better elicit positive reactions from them (p. 416). Further, according to the managers interviewed, the participants improved at getting "the story out . . . a story that economic or financial numbers would support" (p. 416). The writers' ability to find and use evidence to effectively support their ideas and analysis was a significant advancement in the eyes of administrators. Not only did these changes in ability create clearer, more persuasive, more audience-centered writing, they also positively affected the status of the writers within the institution.

In a final workplace study more directly relevant to health-care workplaces, author Lee Spears (1996) reports on a project in which she interviewed 54 nurse managers and 13 nurse educators about their professional writing. Although she acknowledges her sample population was limited, Spears argues that her study "is meaningful because it yielded detailed information from respondents talking about their own experiences" (p. 57). Several of the nurse managers expressed views similar to the following respondent: **"Writing can help you get things done and get what you want. A well-written justification report can help you get new equipment for your department or keep your budget from being cut"** (p. 58).

Spears goes on to report that similar to other workplace writing studies, such as Anderson (1985), Dautermann (1993), as well as Gallion and Kavan (1994), participants in her research cited memos, reports, and proposals as the most frequently written documents. As Anderson discussed in the vignette cited earlier in this chapter, determining the purpose and context for these various documents was a central part of the writing process for the nurse managers. As for issues of audience, Spears' participants listed a wide range of targeted readers. These included "members of the health care team, from physicians to nurses' aides; hospital board members; housekeeping staff; medical equipment suppliers; insurance evaluators; attorneys; and accreditation agencies" (p. 60). In order for the nurse managers to achieve their various communicative goals, it was crucial for them to negotiate the needs and expectations of their diverse audiences.

These scholars illustrate that preparing future professionals to communicate about their work in ways that educate, inform, and persuade their intended audiences is sophisticated work. **Teaching about these issues necessitates providing students with opportunities to develop and cultivate a flexible set of research and rhetorical strategies that can be adapted to always-variable contexts.** Technical and Professional Communication pedagogy that prepares students for the complex work environments they will face must move beyond learning how to use the formats of traditional genres or simply conducting library database research. Because communication that actively persuades others of a particular idea requires credible and compelling evidence, providing students with experiences in gathering information from a variety of primary and secondary sources and using it as effective support for their arguments will more realistically prepare them for the professional challenges they will face.

For educators interested in pushing the boundaries of research, information gathering, and persuasive communication, the course described below offers a number of valuable alternatives to traditional approaches to teaching research methods and the composition of rhetorically savvy communication. While the assignment sequence discussed here certainly includes traditional academic research, this is just one of many components in a set of strategies designed to gain information and understanding to support credible and influential communication. **The intent is not to have students master any one particular method but rather to equip them with a set of tools that can be adapted and used as needed based on the variable rhetorical and contextual situations in which professionals find themselves.**

ASSIGNMENT SEQUENCE: FOCUS ON MULTIPLE-SOURCE RESEARCH FOR CONSTRUCTING PERSUASIVE, EVIDENCE-BASED COMMUNICATION

Introduction

No matter what profession students plan to enter when they graduate, they will face complex communication demands involving multiple audiences, sophisticated rhetorical situations, and ever-changing technologies. It is critical that technical and professional communication courses include attention to audience-centered, persuasive communication that teaches students how to analyze institutional problems, to conduct multiple source research, and to use that work to create persuasive, evidence-based writing.

Because the students enrolled in the course discussed here were pursuing a more advanced degree in nursing, all of them were already working in various health-care positions while going to school. This meant I was able to shape assignments around activities that were relevant and realistic for their actual work environment. My ultimate goal for the course was to have students write a persuasive proposal about a change they would like to see in their workplace. In this advocacy proposal, I wanted them to identify a problem, issue, or conflict, to provide some background on why the problem exists, and to build a convincing and tactful argument addressed to decision makers in their organizations about why the change is both necessary and beneficial. An example of a student's advocacy proposal is available in Appendix C. To prepare students for this culminating project, I developed a sequence of assignments that helped them to analyze the rhetorical situation and to conduct research through a variety of sources so that they would have persuasive, evidence-based support for their proposals. A sample student Research Report that incorporates these primary and secondary sources is available in Appendix B.

Although the course I discuss here was designed specifically to prepare students pursuing RN-BSN degrees in nursing for workplace communication demands, the assignments have value and applicability to a broad population of students in technical, scientific, and socially related disciplines. For example, students in introductory level courses could write about part-time student employment, campus organizations to which they belong; or the course could be based upon a service-learning project for which the following assignments could be tailored.

This sequence of assignments teaches students about analyzing institutional problems, conducting multiple source research, and using that work to create persuasive, evidence-based writing. In each of the course's assignments, students focus on understanding and negotiating the three core components of the rhetorical situation which include

1. **purpose**—the reason for which the writing is done
2. **audience**—the people or groups to whom the communication is directed
3. **context**—the situational conditions in which the text will be read and used

Among the primary learning outcomes I expect from students completing this course are

- **to understand the rhetorical nature of professional writing and that each communication situation requires negotiating the unique audience, purpose, and context for which it is intended.**
- **to demonstrate strategies for learning context-specific and specialized subject matter through conducting individual and collaborative research in a professional setting.**
- **to understand the communicative and social conventions of their field and their workplace and how they influence the planning, development, and reception of communication.**
- **to demonstrate that they understand how to integrate written content, images, and basic design principles in order to create usable, persuasive, and reader-friendly documents.**

Sequencing and Organizing Instruction

The assignment sequence for this course consists of four major projects, each with several substeps (see Appendix A for complete directions for all four assignments). These projects are designed to build upon one another and to prepare students for the challenges of the final proposal.

1. Preproposal Brainstorm

I begin this sequence by asking students to write a 500-word *Preproposal Brainstorm* about an issue of professional interest. They are required to identify an issue relevant to their workplace and briefly discuss how they might like to see it changed. The purpose of this is to generate a focus for the assignments that follow rather than to produce a polished piece of writing. This assignment could easily be modified for a wider variety of majors or disciplines.

2. Express-Line Observation and Meta-Analysis

This assignment has four subparts in which students learn about the value and methodology of qualitative observational research. The first part of this assignment asks students to use the express checkout line at a local grocery store, paying particular attention to the details of their experience. They then write up a 500-word account of their observation and post it to the online course discussion board.

In the second part of this assignment, students read the first chapter from Emerson, Fetz, and Shaw's (1995) text, *Writing Ethnographic Fieldnotes.* In this piece, the authors define the basic purposes and practices of ethnographic research. They also demonstrate the research value of qualitative methodologies, particularly of doing participant observation and of keeping fieldnotes of thick description. Emerson, Fetz, and Shaw write

> Ethnographic field research involves the study of groups and people as they go about their everyday lives. . . . The ethnographer participates in the daily routine of this setting, develops ongoing relations with the people in it, and observes all the while what is going on. . . . [S]econd, the ethnographer writes down in regular, systematic ways what she observes and learns while participating in the daily rounds of life of others. Thus the researcher creates an accumulating written record of these observations and experiences (p. 1).

Importantly, Emerson, Fetz, and Shaw also go on to show how, in three examples of express-line observations written by their own students, people's accounts of similar experiences always differ from one another's because of individual backgrounds, biases, and ways of seeing. They point out that while it is important for ethnographic researchers to acknowledge their own partiality, their observations and understandings nonetheless have value in understanding a particular context. They explain that

> Writing fieldnote descriptions, then, is not so much a matter of passively copying down "facts" about "what happened." Rather, **such writing involves active processes of interpretation and sense-making: noting and writing down some things as "significant," noting but ignoring others as "not significant," and even missing other possibly significant things altogether (p. 8).**

This last point is particularly important for students trained in a scientific, quantitative-based discipline as they have often not had experience with this methodology.

The third part of this assignment has students read at least two observations done by their classmates and to write a 650 word reflection and meta-analysis that ties together the reading, their own observation, and the observations recorded by their classmates. These analyses are particularly interesting because students begin to see value in different ways of gathering information, though they are not necessarily convinced that others will see this "evidence" in the same light.

Finally, students are asked to conduct a short qualitative observation in their workplace in which they focus on interaction and activities related to the issue identified in their Preproposal. Using what they learned through their interpretation of this experience and their written field notes, students use this information later in their Research Report and their Final Proposal.

3. Research Report on Issue of Professional Interest
Utilizing Multiple Primary and Secondary Sources

This assignment asks students to write a five page report that identifies and discusses perspectives related to the workplace issue identified in their initial Preproposal. Using multiple primary and secondary sources, students should classify and investigate various views on their subject, as well as how their respective proponents argue for them. Beyond helping them to investigate the issue, this work is intended to prepare students to provide convincing support for the value of the changes they will suggest in their proposals.

In addition to including the qualitative observation and visual/textual analysis done in the previous assignments, this step in the sequence also requires students to conduct at least one formal interview with a co-worker (though I encourage them to interview several people). Because any change in workplace policies or procedures needs the support of co-workers, collaboration is a crucial step. One way to develop collaboration and to learn about the concerns and needs of others is through conducting interviews. As part of this assignment, I have students do a set of readings on interviews and we discuss strategies for interacting with co-workers in a way that encourages collaboration and elaboration rather than defensiveness.

Additionally, as part of this assignment, students are required to find at least two academic articles examining their issue. The inclusion of published, scholarly research not only helps writers better understand the issue about which they are writing, but also helps them to present a stronger, more persuasive, more credible case to their target audience.

Finally, using the standard conventions for reports, students write up their findings to help readers understand the global and local intricacies of their issues. These findings will form the evidentiary basis for the final assignment in the sequence: the Professional Advocacy Proposal for Workplace Change. A sample student Research Report is available in **Appendix B.**

4. Professional Advocacy Proposal for Workplace Change

With ample background research completed and the issue analyzed from multiple perspectives, students return to their preproposal idea for the final assignment. In this final step, they are asked to write a persuasive, evidence-based proposal that advocates a change of policy, procedure, or activity in their workplace. In addition to observations, textual analyses, workplace observations, and the results from their interviews with co-workers, students use academic research articles to provide support and justification for their proposed ideas. After drafting a version of their proposal, they receive feedback from me, a pair of peers, and, if possible, one of their co-workers or a supervisor.

Assessment

Assessment of the final revised version of the Advocacy Proposal is based on how well students address the primary goals of the assignment. Specifically, my evaluation examines the writers' ability to

- **introduce** their subject and provide contextualization of the workplace setting and issue
- **identify** the sources of the problem, conflict, or inefficiency that the proposal addresses
- **argue for and support** their proposed idea in a way that addresses the concerns of all stakeholders in the issue
- **structure and organize** their document so that it facilitates clarity and ease of use

An example of a student's advocacy proposal is available in **Appendix C.**

CONCLUSION:
DISCUSSION OF OUTCOMES FROM THE COURSE

The use of multiple primary and secondary sources for research has proved very productive for the work produced by students. Preparing and requiring students to do observations in their own workplaces is useful in helping them to see activities and interactions that are likely implicit in their own everyday work. By making the familiar strange, students are observant of people and practices they might otherwise miss. One successful observation was done by a student who works for a county health department. With permission, she observed three teenage girls seeking information from a fellow health department nurse about birth control and contraception. Being in the position of a participant, rather than in her usual role of working with patients, this student was able to gather a number of important insights about the needs of this particular population. They mentioned their difficulty in finding transportation from their rural community to the clinic and in getting an appointment in the late afternoon so that they would not have to miss school. This information, in turn, helped the student to make a stronger case for providing family planning services after school hours.

In analyzing internal institutional documents, another student chose to look at a publicly available evaluation report of the nursing-home facility where she worked in order to find statistics on medication and treatment errors, as well as injuries to staff. In a proposal arguing for reducing staff-to-patient ratios and staff overtime, she was able to show that 7 of 13 critical errors cited in the report occurred when the nurse had been on duty greater than 14 hours.

In conducting interviews with co-workers, one student working in a long-term care facility was encouraged by her peer reviewers to seek out the patient Ombudsman as a way of getting input on proposed changes from patients. Doing so helped her to pay attention to a more comprehensive set of stake-holders in her bid to reassign nursing and supervisory duties in each wing of the facility.

Finally, using published, peer-reviewed research, students were able to reference credible evidence that was persuasive to both peers and administrators. One student used a study on the effects of noise on newborns in her proposal to help prove to her co-workers that there is a legitimate need to eliminate unnecessary noise in the neo-natal intensive-care unit where she works. Although many of her co-workers agreed with her idea, they had been at a loss as to how to make reductions and how to determine what levels of noise were acceptable. This research provided a definitive guideline for shaping the new proposed policy.

In each of these instances, students were able to select from a rich array of information to build the case for their advocacy proposals. **By conducting research from diverse sources, students were in a position to choose the most persuasive details for their given audiences, purposes, and contexts. Rather than relying solely on traditional academic research, this sequence of assignments helped students to discover that credible and compelling communication is most often achieved through utilizing multiple modes and sources.**

Because of complex professional working environments, nurses need persuasive communication practices that will provide them with a credible voice in the activities, policies, and procedures of their workplaces. However, learning to construct influential communication using multiple primary and secondary sources of support will help students from all disciplines to develop professionally, to provide better advocacy for their co-workers and clients, and to be more integral and productive members of the organizations where they work. As instructors of professional and technical communication, we must provide our students with opportunities for developing a rich array of persuasive, research-based rhetorical practices so that they may learn to address successfully the diverse audience, purposes, and contexts in which they will find themselves as professionals.

APPENDIX A
Using Multiple-Source Research to Create Persuasive, Evidence-Based Communication

ASSIGNMENT SEQUENCE DIRECTIONS

1. Pre-Proposal Brainstorm

Begin by reflecting on your response to Cipriano's chapter "Contemporary Issues in the Health-Care Workplace" (2002, chapter 15) in your Mason, Leavitt, and Chaffee textbook, as well as problems or inefficiencies you see on the job. Select an issue of professional interest that is relevant in your workplace. Write an informal, 500-word brainstorm about the issue and how you might like to see it changed within your workplace. Make sure that the issue has at least some component that you can realistically argue about for change. The purpose of this step of the assignment is to generate ideas and possibilities, not to produce a polished piece of writing. The more you are personally invested in this issue, the easier it will be to write your proposal. When you are finished, e-mail your brainstorm to me as a Microsoft Word (.doc) or Rich Text Format (.rtf) attachment.

2. Grocery Express Line Observation Assignment

Part 1: Observation

- Use the express checkout line at a local grocery store of your choice. Pay particular attention to the details of your experience. Immediately following, write a 500-word observation of your experience.
- Because there are no right or wrong approaches or answers to this assignment, I'm not going to give you more direction than above about what to do.
- Post your observation on the WebCT discussion board under this assignment's title.

Part 2: Meta-Analysis

- Once you have posted your Grocery Express-Line Observation to the course Discussion thread, read Emerson, Fetz, and Shaw's article, "Writing Ethnographic Fieldnotes" (1995).
- After reading the article, go to the Discussion link and read at least two Grocery Express-Line Observations by other classmates.
- Write a 650-word reflection and meta-analysis that ties together the reading, your observation, and your classmates' observations. Your reflection and meta-analysis should address the following questions:
  —What did you learn from this experience?
  —How did your approach differ from that of your classmates and from how a professional ethnographic researcher might approach such work?
  —What kinds of similarities and differences existed?
  —How might participant-observation help you to better understand activities in your work setting?

Post your response on the meta-analysis discussion thread on WebCT.

3. Research Report on Issue of Professional Interest Using Observations, Interviews, and Published Resources

Purpose

This assignment has three main goals. First, gathering research from a variety of sources will help you to find credible evidence to support the change you will argue for in your proposal. Second, this assignment will give you practice in collecting primary-source research through close observation and interviewing your co-workers. Third, through searching for secondary source research, such as medical journals and online databases, you will gain experience finding, presenting, and citing secondary-source research in your writing.

Directions

1. *Read Burnett's* chapter (from *Technical Communication, 5th edition* (2001)) "Locating and Recording Information," which discusses practical strategies for gathering research.

2. *Write up a series of questions and interview at least one co-worker.*
   Using the guidelines from pages 172–176 in Burnett, write a short set of questions to use in interviewing at least one co-worker about the issue you will discuss in your proposal. Be sure that you write/ask your questions in an open-ended way so that your respondent(s) can answer in more depth than simply giving you a yes or no answer. You will need to write up your questions and what you discovered from this interview to be turned in prior to the draft of your report.

3. *Conduct a short observation in your workplace.*
   Using what you learned from the Express-Line Assignment and reading, as well as Burnett's discussion on pages 168–170, conduct a short observation in your workplace. Concentrate directly on activities, interactions, and details that will help you to argue for change in your proposal. Try to record specific examples or details that will illustrate your observations. Remember that when you are writing up your observation, it is critical to identify any biases that may influence your views and to identify the context of what you saw for your readers.

4. *Find and analyze at least two published articles relevant to your issue.*
   Using your professional publications and the databases available to you through the library or your workplace, **locate at least two academic articles** relevant to your issue. Since you will be using this work to support your proposal's argument, you should try to find sources that take diverse perspectives so you will be aware of the subissues you'll need to address. Nearly all of the articles in your Mason, Leavitt, and Chaffee textbook cite academic sources. Use their bibliographies as leads for finding additional sources.

5. *Write a draft report.*
   Your 5-page/1250-word report should identify and discuss the perspectives on the issue you found through your research. Beyond helping you to investigate your issue, this work will prepare you for providing convincing support for the value of the change you will suggest in your proposal.

   Report structure
   You should use the following format to present your findings:
   - **Introduction:** What issue are you addressing and what will readers gain from reading your report?
   - **Method and Facts Subsections:** Using a subheading to separate them, discuss each method you used for gathering information and what you found through this research.
   - **Discussion and Conclusion:** What does the material you gathered mean? Why is it credible or believable? Why is it significant or important?
   - **References:** Be sure to include a list of references used at the end of your paper. It should be formatted in either APA or MLA style.
   - **Appendix:** Include your interview questions and any other relevant material.

6. *Give feedback and suggestions to assigned classmate's Research Report.*
   Look at the WebCT Discussion thread for "Draft Reports"; read and provide a thoughtful critique for your assigned classmate's report using the reply feature.

7. *Revise draft Research Report.*
Addressing the commentary and suggestions given by your classmate and me, revise and polish your Research Report into a final document.

4. *Professional Advocacy Proposal for Workplace Change Assignment Directions*

Purpose

The steps in this project are designed to pull together the activities and work you will do in other course assignments. Your ultimate task is to write a persuasive proposal about a change you would like to see in your workplace. You will begin by identifying a problem, issue, or conflict and providing some background on why this problem exists. You will then construct a convincing and tactful argument about why this change is both necessary and beneficial. Additionally, you will use workplace observations, informal interviews with co-workers, and academic research articles to provide support and justification for your proposal. **You should view your intended audience as the people, committee(s), or governing body with the authority to enact your proposal.**

Directions

In addition to the Preproposal brainstorm you completed at the beginning of the semester, this assignment has three parts:

1. *Proposal Draft*
Once you have received feedback from me about your Proposal subject, begin work on the actual document. Minimally, your draft should provide some background and context for readers, a clear explanation of what you would like to see changed, and some research support that helps to justify your plan. Post your draft to the WebCT Discussion thread entitled "Draft Proposals."
2. *Feedback and Suggestions to Assigned Classmate's Proposals*
Read and provide thoughtful critique for your assigned classmate's proposal using the reply feature in WebCT.
3. *Final Proposal*
After receiving feedback and suggestions on your Draft Proposal, revise your document so that it is polished, professional, and persuasive. Your Proposal will be assessed based on how well you
   * **introduce** your subject and provide contextualization of the workplace setting and issue;
   * **identify** the sources of conflict or inefficiency that your proposal addresses;
   * **argue for and support** your proposed idea in a way that addresses the concerns other stakeholders in the issue may have;
   * **structure and organize** your document so that it facilitates clarity and ease of use.

APPENDIX B
Using Multiple-Source Research to Create Persuasive, Evidence-Based Communication

Sample Student Research Report on Issue of Professional Interest

Decreasing Nurse Initiated Pacifier Use in the Newborn Nursery

Submitted by Joan Eanes
New Mexico State University

Introduction

It is common knowledge that breastfeeding is the optimum choice in feeding a newborn for many reasons. This is supported by the American Association of Pediatrics, the World Health Organization, and the National Healthy People 2010 objectives and backed by years of research. I have found that most women make the decision to breast or bottle feed before the last trimester of their pregnancy. When they come to the hospital it is not our job to "make" them breastfeed, but to support their choice. When a woman makes the decision to breastfeed, many factors effect how long it will take the mother and infant to "get it right". How nurses interact with the infants, and what messages we give, through verbal communication and modeled behavior, to the moms can directly effect how easily the transition to successful breastfeeding occurs. Pacifier use and especially nurse initiated pacifier use can have a negative effect on the breastfeeding process. When a nurse initiates newborn pacifier use it should be based on research and policy. There are medical indications for pacifier use, such as during tube feeding to better increases gut transit time for premature infants, and Carbajal, Chauvet and Couderc, et al (1999) have demonstrated that pacifiers have been shown to be more effective than sweet solutions for newborn pain management during painful procedures (p. 1393-1397). But when used without medical justification, pacifiers can have other unwanted effects. Newborns are born with a set of reflexes to assist with learning to breastfeed, but how stressful the birth, what medications, and other interventions they are exposed to during the birthing process can have a strong effect on how easily they transition to full breastfeeding. Breastfeeding is a learned activity for the baby and the mother, taking time and coordination not unlike learning to dance with a partner. Feeding with a bottle and an artificial nipple or using a pacifier involves different motor skills and stimulation than breastfeeding. For those infants behind the power curve to start with due to the birth experience, medication exposure, or slight prematurity, early use of a pacifier can further interfere with the infant's ability to initiate and establish breastfeeding and long term pacifier use has further implications for maintaining milk supply, infection, and dental problems. On our unit, nurses are supportive of breastfeeding, but like everyone else have strong opinions affected by our culture, their own experiences, time constraints and stressors, and occasionally initiate pacifier use without understanding the underlying effects a seemingly inconsequential decision may have on an individual infant and what message we are giving the infant's family. I hope to explore some of the evidence supporting our breast-feeding policy's directive on pacifier use.

Method and Facts
Interviews:

I did two phone interviews and three face-to-face interviews with three nurses from the day shift and two nurses from the night shift. They have Maternal Child nursing experience ranging from one to 15 years. I asked a series of the same six questions of each of the nurses. The questions were designed to get a sense of the nurse's policy and general knowledge base, and opinions concerning using pacifiers with newborns initiating breastfeeding. Three nurses were familiar with our unit's breastfeeding policy statements concerning pacifier use. A copy of the policy was available for review for those not familiar with the policy. The Maternal Child Unit's policy on breastfeeding states: "Pacifiers should be avoided until the baby has established an adequate breastfeeding pattern unless requested by the mother after being informed of the risk factors involved".

The nurses interviewed felt that 90 to 100 percent of the time they followed policy, but they felt that other nurses on the unit followed the policy only 50 – 90 percent of the time. The factors they identified as affecting their clinical practice included: the policy, research, that the unit does not stock pacifiers, their own breastfeeding experiences, how they were originally trained and the parents' attitudes.

I asked for suggestions on how to help the staff better understand and follow the Breastfeeding Policy Directive on pacifier use. Some of the suggestions from the nurses included providing additional, research-based training, and listing risk factors in the policy. Some felt that those that didn't follow the policy should be "written-up". They also suggested that the Lactation Educator should follow-up with any staff member that inappropriately initiates a pacifier. The Lactation Educator would then have an opportunity to provide education if needed and to identify those staff members that continue to have a problem following the policy and then to pass that information on to the Director of the nursing unit for disciplinary action if needed.

Observations:

My observations in our Newborn Nursery included finding a pacifier in a premature infant's crib and in the crib of a term infant having breastfeeding problems. There was no documentation in the infant's chart whether it was nurse or parent initiated, of education of the mother about the risk factors involved in early pacifier use, or for the rational for the use. The documentation in the patient's chart is the communication tool of the professionals involved in that patient's care. Since our policy, as previously mentioned, gives parameters for the use of a pacifier and for the education of the parents when they request initiation, I would have expected to see some justification for the pacifier, and what information was given to the parents.

I worked as the Lactation Consultant on our unit for several years. Most of the nurses know my opinion of pacifier use and are a little defensive about the topic or want the "other nurse" held accountable. They verbalize that sometimes it is very busy and a mother will send a crying baby to the nursery when she is not willing or able to put the infant to breast. If they don't have the time to hold the infant they feel frustrated, don't want the infant to suffer because the mother won't do what they perceive she should be doing and will give a pacifier to quiet the baby. Usually what I hear from the nurse on one shift is that the nurse from the opposite shift gave the infant the pacifier and "someone should talk to her about that". The present Lactation Educator says she now better understands my views after working with mothers and infants with breastfeeding difficulties and finding so many of them using pacifiers.

Secondary research:
I used Marie Biancuzzo's *Breastfeeding the Newborn, Clinical Strategies for Nurses (2003)* as a resource for research and evidence-based practice suggestions. Other sources of information include research articles from various journals, policy state-ments from the American Association of Pediatricians, and the United Nations Children's Fund (UNICEF) and the World Health Organization (WHO) recom-mendations. I discuss these in the following section.

<u>Discussion</u>
Our Maternal Child Unit Breastfeeding policy is based on *The Baby-Friendly Hospital Initiative's Ten Steps to Successful Breastfeeding for Hospitals* (1998) developed by the United Nations Children's Fund (UNICEF) and the World Health Organization (WHO). In 1989 WHO/UNICEF put forth their statement *Protecting, Promoting, and Supporting Breastfeeding: The Special Role of Maternity Services.* Marie Biancuzzo (2003) points out that:

This statement, which outlined universally relevant principles and action steps, was intended as a summary of what needed to be done to improve perinatal breast-feeding efforts throughout the world. The 'Ten Steps to Successful Breastfeeding' contained in that document was intended as an executive summary and later became the cornerstone for the Baby-Friendly Hospital Initiative (BFHI) in 1991. (Biancuzzo, p. 210)

The ninth step as outlined for the United States is as follows: "Give no artificial teats, pacifiers, dummies, or soothers to breastfeeding infants" (Biancuzzo, 2003, p. 217). Our policy states that "Pacifiers should be avoided until the baby has established an adequate breastfeeding pattern unless requested by the mother after being informed of the risk factors involved". Marie Biancuzzo (2003) identifies a primary barrier to reducing or eliminating pacifier use is that "parents and health care providers believe that they are helpful or at least harmless for healthy term newborns" (p. 217).

Aside from the medical indications discussed previously, pacifiers can have a negative effect on the infant. How often and how effectively the infant goes to breast in the first weeks of life affects the transfer and establishment of the milk supply. Milk supply is basically established and maintained by how well and how often the infant breast-feeds. The amount of milk transferred from the breast tells the mother's body how much will be needed at the next feeding. How well the infant transfers milk from the mother is a process of efficient latch on, and the stimulation of mother's hormones to facilitate "let down" and the release of milk. A mother has colostrums available for the baby from birth. The regular milk supply comes in usually two to three days after birth and is directly affected by the effectiveness of the infant's latch on and the frequency of the feeds. Biancuzzo (2003) points out that "it has been clearly established that the mechanism used to suck an artificial nipple is different from that used to suck a human nipple" (p. 210). This, in turn, suggests "that it is difficult for some infants to attach to the mother's nipple after they have learned to suck the artificial nipple" (Biancuzzo, 2003, p. 210), and that "imprinting on artificial objects can occur" (Biancuzzo, 2003, p. 210). Mothers and infants are usually in the hospital for one to three days. When pacifiers are introduced during the first few days of life it is difficult to predict which infants will have problems with latch-on and how it will affect the whole process. Infants who attach to the breast using the same technique used with a pacifier can have poor milk transfer, and cause their mothers' pain from improper latch on, which starts a cycle of the hungry baby wanting to go to breast more often due to the poor milk transfer, a mother that is disinclined to want that infant at the breast more frequently due to pain and a higher use of the pacifier. Biancuzzo (2003) cites, Centuori and colleagues and their observation "that pacifier use during hospitalization is significantly related to nipple soreness during that time" (p. 211). As nurses we have a duty not to contribute to this cycle. I don't think nurses intentionally want to interfere, but with early discharge (usually one to three days after delivery), we don't see the consequences of our actions.

Biancuzzo (2003) writes that pacifiers have been associated with decreased frequency of feedings and decreased frequency of feedings can lead to the establishment of an inadequate milk supply and to decreasing an already established milk supply (p. 211). I have heard nurses tell mother's to use a pacifier based on the perception that the infant is "using you like a pacifier". Infants do have a high sucking need, but parents and healthcare providers "express little or no acknowledgment that suckling the breast meets that need" (Biancuzzo, 2003, p. 211). Breastfeeding frequency should be based on infant feeding cues and infant led. We do the parents a disservice when we don't help then to learn other quieting and soothing techniques for those times when breastfeeding doesn't seem to be the answer.

CONCLUSION

It can take from several days to several weeks to achieve effective, comfortable breastfeeding. With the hospital stay as short as 24 to 48 hours for an uncomplicated delivery, helping a new mother get a good start on breastfeeding is sometimes difficult in the short amount of time available. Even though pacifiers seem to be ever-present in our society, as nurses we need to recognize the negative effects they can have during the precious few hours we do have to help a mother and infant succeed in breastfeeding. On our Maternal Child Unit we have a policy that states "Pacifiers should be avoided until the baby has established an adequate breast-feeding pattern unless requested by the mother after being informed of the risk factors involved". The Association of Women's Health, Obstetric, and Neonatal Nurses strongly supports evidence-based practice and our policy is based on evidence and research. We need to be aware of the research available concerning pacifier use and let that lead our practice. Infants need to have the chance to establish breastfeeding without our interference. By not initiating pacifier use during the critical first days of the breastfeeding process we can help infants to have one less impedance to learning to latch-on, to transfer milk effectively, and to breastfeed without causing pain to their mothers. Infants are exposed to many factors that can effect how well they are able to breastfeed, but one way we can help to facilitate the process is to follow our unit policy on pacifier use and breast.

References

American Academy of Pediatrics. (1982). The promotion of breastfeeding. Policy statement based on task force report. *Pediatrics* 69, 654-661.

Biancuzzo, Marie. (2003). Hospital practices to support successful breastfeeding. In *Breast-feeding the Newborn, Clinical Strategies for Nurses* (pp. 181-222). St. Louis: Mosby.

Carbajal, R., Chauvet, X., Couderc, S. et al. (1999). Randomized trial of analgesic effects of sucrose, glucose, and pacifiers in term neonates. *BMJ* 319, 1393-1397.

Centuori, S., Burmaz, T., Ronfani, L., et al. (1999). Nipple care, sore nipples, and breast-feeding: a randomized trial. *Journal Human Lactation* 15, 125-130.

World Health Organization and United Nations Children's Fund. (1989) Protecting, pro-moting, and supporting breast-feeding: the special role of maternity services. Geneva, Switzerland: World Health Organization.

World Health Organization. (1998). *Evidence for the ten steps to successful breastfeed.* Geneva, Switzerland: World Health Organization.

APPENDIX C
Using Multiple-Source Research to Create Persuasive, Evidence-Based Communication

**Sample Student Professional Advocacy Proposal
for Workplace Change**

PROPOSED:
COMPUTER NETWORKING SYSTEM

Submitted to
Dr. Tom Lindsey, CEO Mimbres Surgical Consultants
Dr. Joel Jones, CEO Blue Skies Primary Care
Derrick Yu, CEO Mimbres Memorial Hospital

Prepared by
Mary Frosch, RN
Mimbres Surgical Consultants
Mimbres Memorial Hospital

Proposal Abstract

A computerized system networking the departments of Mimbres Surgical Consultants and interfacing with Mimbres Memorial Hospital will improve efficiency and reduce patient wait times. An educational program using computerized data can also reduce patient anxiety and yield greater understanding. Valuable time will be saved through the efficiency of this system and the benefits will outweigh the initial costs of updating the current system and computer program. As the costs of providing medical care increases and reimbursements decrease, an effort must be made by physicians to see more patients over a lesser period of time.

Introduction

As part of a team of Mimbres Surgical Consultants, we all feel it is important to maintain patient satisfaction through efficient use of our time and resources.

A computerized system networking the departments of Mimbres Surgical Consultants will increase efficiency in the workplace. This system will be able to interface with Mimbres Memorial Hospital as well, reducing time and manpower required to access information at both facilities. This proposal also calls for an educational program, utilizing computerized data, for providing patients with updated information in an environment conducive to learning. This may help to decrease the patient's anxiety by yielding patient understanding and satisfaction.

History

The current computer system, which was implemented one and one-half years ago, interfaces between the receptionist and the accounting departments only. Neither physicians have computers in their offices nor do the nursing areas have access to a computer system. The laboratory has a DOS-based system that does not interface with the current system, which is a Windows-based management system. Delays in patient evaluations by physicians have occurred due to the lack of information from the hospital's Medical Records Department. Telephone calls requesting patient information have taken hospital employees away from the transcription of information, which in turn has caused a backlog of up to two weeks for physician's to receive dictated information. This in turn has affected the hospital financially, as the need for contracting with an outside transcribing company has become necessary. Interfacing electronically with Mimbres Memorial Hospital will allow each specific doctor the ability to access his patient's information promptly and efficiently. As the community grows, so must our ability to see more patients in the same amount of time currently spent.

The same is true with patient education. Not all people have access to current technology, and are unable to research their conditions on the Internet. By accessing a Web-based educational system, all patients will be able to review current medical information regarding their disease or intended procedure/surgery. Providing adequate information to patients will allow them to make an informed choice and decision about their particular condition.

Plan
The plan for a networking system between the Mimbres Surgical Consultants and Mimbres Memorial Hospital depends largely on collaboration. Each facility must evaluate and determine how this can be done in a timely and cost efficient manner. The most crucial element is that data flow back and forth freely. The systems must interface properly, thus avoiding the need to punch data redundantly during appointments, accounting and billing.

Benefits
The primary benefit of this proposal is increased employee efficiency at both facilities. Additionally, by decreasing the amount of time spent in obtaining vital information, the patient and physician satisfaction will improve. Another benefit of this proposal will be improved individualized patient education, which will relieve anxiety and increase patient satisfaction.

Approach
Methodology
The proposed plan for the implementation of the Networking System will be accomplished in an orderly and efficient manner, as to maintain quality of care at both institutions. The plan is as follows:

• An evaluation of the current systems by each facility's computer technologist will need to be completed prior to the purchase of any new equipment or software. Each facility's technologists will need to meet with the CEO's from the doctor's office as well as the hospital.

- After the evaluation is complete, a decision agreeable to both facilities regarding software and hardware purchases may then be made. Lytec is the program currently being used at Mimbres Surgical Consultants. This software has the capability for services other than what is currently being used. The decision to purchase a computer for each Physician's office will be at the discretion of each physician.

- A request may then be submitted to the computer company of each facility's choice; the current company used by Mimbres Surgical Consultant's is Dell. Dell currently has a Microsoft Office System comprised of simple to use programs, servers, and services that helps ensure people and organizations stay connected to information, business processes, and each other.

- Computer terminals will need to be placed at the nurses' station. There is a computer terminal with internet access in the office that could be used for patient education.

- Internet access will need to be available at each nursing computer station. An alternative to consider is the use of laptops and a wireless system, which will allow for portability and far more flexible when scheduling patients.

- Internet education systems may be reviewed by all physicians, so that the educational material will be pertinent to each physician's practice.

- After the installation of the computer system, staff education on its use will be provided by a designated representative. The best time to in-service staff would be after patients are seen, in the evenings. An alternative time would be on a weekend, which would impact both institutions financially.

- Evaluation of the system shall be done one month after implementation.

- Any additional suggestions for changes or upgrades may be submitted in writing by staff members at any time.

Schedule

The estimated time for completion of the proposal, for Mimbres Surgical Consultants, will be six weeks after the initial evaluation by the computer technologists. The time may be subject to change due to the availability of necessary equipment, software, and hardware.

Costs/Budget

The estimated budget for this proposal will be determined by the CEO's of Mimbres Surgical Consultants and Blue Skies Primary Care once the needs have been assessed. The budget will need to include pricing for a computer and printer for the nurses' station, a computer for the designated patient education area, as well as a server to interface with our existing system.

The budget may be reduced by the following:
- Education of staff to the new system is done by an in-house individual.

- Keep the existing Lytec system for scheduling and billing programs.
- Use the existing computers in the reception and accounting departments.
- Use the empty physician office for the patient education area, which is already furnished and has internet access.
- Use web sites for education that do not require additional funds to access.

Evaluation

A computerized system would be beneficial to Mimbres Surgical Consultants. Not only would this be beneficial to patient care, but would also facilitate efficiency in the workplace. Through the development of this system, an increase in patient satisfaction and the efficiency of nursing staff should be apparent.

REFERENCES

Anderson, P. V. (2003). Introduction. *Technical communication: A reader-centered approach* (5th ed.). Boston, MA: Thompson/Heinle.

Anderson, P. V. (1985). What survey research tells us about writing at work. In L. Odell & D. Goswami (Eds.), *Writing in nonacademic settings* (pp. 3-83). New York: Guilford Press.

Burnett, R. (2001). *Technical communication* (5th ed.; pp. 161-206). Boston, MA: Thompson/Heinle.

Chaffee, M. W. (2002). Communication skills for political success. In D. Mason, J. Leavitt, & M. Chaffee (Eds.), *Policy and politics in nursing and health care* (4th ed.; pp. 311-332). St. Louis, MO: Elsevier Science.

Cipriano, P. F. (2002). Contemporary issues in the health care workplace. In D. Mason, J. Leavitt, & M. Chaffee (Eds.), *Policy and politics in nursing and health care* (4th ed.; pp. 93-111). St. Louis, MO: Elsevier Science.

Dautermann, J. (1993). Negotiating meaning in a hospital discourse community. In R. Spilka (Ed.), *Writing in the workplace* (pp. 98-110). Carbondale, IL: Southern Illinois University Press.

Emerson, R., Fetz R., & Shaw, L. (1995). *Writing ethnographic fieldnotes.* Chicago, IL: University of Chicago Press.

Gallion, L. M., & Kavan, C. B. (1994). A case study in business writing: An examination of documents written by executives and managers. *Bulletin of the Association for Business Communication, 57*(4), 9-11.

Leavitt, J. K., Cohen, S. S., & Mason, D. J. (2002). Political analysis and strategies. In D. Mason, J. Leavitt, & M. Chaffee (Eds.), *Policy and politics in nursing and health care* (4th ed.; pp. 71-91). St. Louis, MO: Elsevier Science.

MacKinnon, J. (2003). Becoming a rhetor: Developing writing ability in a mature, writing-intensive organization. In T. Peeples (Ed.), *Professional writing and rhetoric: Readings from the field* (pp. 411-422). New York: Longman.

National Commission on Writing for America's Families, Schools, and Colleges. (2004). *Writing: A ticket to work . . . Or a ticket out: A survey of business leaders.* Retrieved January 27, 2005 from
http://www.writingcommission.org/prod_downloads/writingcom/writing-ticket-to-work.pdf

Paré, A. (2002). Keeping writing in its place: A participatory action approach to workplace communication. In B. Mirel & R. Spilka (Eds.), *Reshaping technical communication: New directions and challenges for the 21st century* (pp. 57-73). Mahwah, NJ: Lawrence Erlbaum Associates.

Spears, L. A. (1996). The writing of nurse managers: A neglected area of professional communication research. *Business Communication Quarterly, 59*(1) 54-68.

CHAPTER 4

Using Design Approaches to Help Students Develop Engaging and Effective Materials that Teach Scientific and Technical Concepts

Anne Frances Wysocki

> **OUTCOME:** Students will understand how and why to use design approaches in developing technical and scientific communications and will be able to use different modalities rhetorically in developing communications.

WHAT BUSINESS AND INDUSTRY TELL US

Design processes and approaches permeate the technical and engineering world—with emphasis on effective communication. Engineers recognize the importance of design practices because engineers make things—bridges, handtools, circuits, diapers—that must work within the complex contexts of human activity. As they research who will use the things they make and when, where, and why those people will use them, engineers must communicate carefully with others. They must also understand and continue to learn about the materials with which they build. In these design processes, it is impossible to separate design from either material or communication—as the authors of a report on "integrating professional practice" into undergraduate engineering classes describe when they give this definition of design:

> the complete design process includes identifying a need or problem, recognizing constraints, identifying and developing courses of action, testing potential courses of action, selecting optimum courses of action, preparing the documents required for the design, managing the overall process, communicating the design, construction, and testing (Butkus & Kelley, 2004, p. 166).

63

Design approaches can be used across disciplines—because people who think as designers function well in the new work environments. While chair of the National Endowment for the Arts, Jane Alexander wrote that

> The "knowledge worker" now in demand is a person who works well in a team, particularly with people from different cultural backgrounds. Such workers also know how to access, evaluate, interpret, and communicate information in a variety of media. They have the curiosity and creativity to pose questions and to innovate. They can grasp the dynamic relationships among parts that constitute larger systems (Davis et al., 1997, p. vi).

The words are from the introduction to a book describing a research project that studied how teachers might incorporate (as the authors of the book put it) a "designerly way of thinking, knowing, and doing" (p. ix); the project's goal was to help students reach the goals of the Secretary's Commission on Achieving Necessary Skills (SCANS). Alexander defines "design" as a process in which students "identify needs, frame problems, work collaboratively, explore and appreciate the contexts in which the solution must work, weigh alternatives, and communicate their ideas verbally, graphically, and in three dimensions" (p. vi). The book from which this comes, *Design as a Catalyst for Learning*, argues that learning this process can prepare students in any field to be the "knowledge workers" required by changing workplaces.

Such definitions of design similarly answer calls for changes in education put forth by the business and education stakeholders (including the U.S. Department of Education, the National Education Association, Apple, Dell, Microsoft, and Cisco, among others) of the "Partnership for 21st Century Skills." In this group's 2003 report, "Learning for the 21st Century" (2003a), and in their specific recommendations to educators and policymakers in "The Road to 21st Century Learning" (2003b), the group emphasizes the skills all students need:

- Information and media literacy skills: Analyzing, accessing, managing, integrating, evaluating, and creating information in a variety of forms and media. . . .
- Communication skills: Understanding, managing, and creating effective oral, written, and multimedia communication in a variety of forms and contexts. . . .
- Problem identification, formulation, and solution: Ability to frame, analyze, and solve problems.
- Creativity and intellectual curiosity: Developing, implementing, and communicating new ideas to others, staying open and responsive to new and diverse perspectives.
- Interpersonal and collaborative skills: Demonstrating teamwork and leadership; adapting to varied roles and responsibilities, working productively with others . . . (Partnership for 21st Century Skills, 2003a, p. 9).

Design practices—as the definitions of "design" I have quoted show— incorporate such skills: they can help students move and act effectively within workplaces that no longer offer the fixed positions or standardized jobs of the last century.

Design processes can help students develop technical communication that works particularly well within its context. Imagine that a team of technical communicators is charged with helping a lay audience understand neighborhood pollution levels following a manufacturing plant's decommision. For such purposes, Natalie Jeremijenko, an artist/scientist who has taught at Yale, NYU, and the University of California at San Diego, worked with undergraduates and high school students to retool inexpensive toy dog robots into "Feral Robotic Dogs" (see Figure 1). The dogs work in packs to sniff out contaminants; as the robots move over a site, they

> provide information that is displayed in a form that is legible to diverse participants. . . . The dog paths provide immediate imagery to sustain discussion and interpretation of an otherwise imperceptible environmental condition of interest. . . . Because the dogs' space-filling logic emulates a familiar behavior, i.e. they appear to be "sniffing something out," participants can watch and try to make sense of this data without the technical or scientific training required to be comfortable interpreting an EPA document on the same material ("Project Missions").

The feral robot dogs are as unexpected an example of technical communication as they are effective: they do what they are supposed to do—help their audience see and understand pollution levels—because the dogs result from a design process involving a broad and careful exploration of purpose, audience, context, and the materials of communication. Because a design approach to technical communication emphasizes creative exploration, including being open to possibilities of different kinds of communication than what might be expected, it encourages solutions that have as close a fit as possible with the particularities of the context. Most technical communication students will work in situations where text is the preferred method of communication; using design processes for developing communication and considering alternative forms of communication, however, can strengthen students' ability to produce communications that work precisely because they play both with and against expectations.

WHAT ACADEMIC RESEARCH TELL US

Recognizing the changing world of work and communication, literacy researchers recognize the need for design approaches. Gunther Kress argues that we are at a "moment in the long history of writing when four momentous changes are taking place simultaneously: social, economic, communicational, and technological" (2003, p. 9). When conditions are stable, Kress believes we ought to teach students "critique," in which "existing forms, and the social relations of which they are manifestations, are subjected to a distanced, analytical scrutiny to reveal the rules of their constitution" (1999, p. 87); given current changes, however, he argues that

Figure 1. The photograph shows the feral robotic dogs at work in the Bronx.
Source: http://xdesign.ucsd.edu/%7Eids/index.cgi?mode=image&album=/
feral%20bronx&image=bronx2.jpg)

in the context of the multi-modal, multi-media modes of textual production in the era of electronic technologies, the task of text-makers is that of complex orchestration. Further, individuals are now seen as the remakers, transformers, of sets of representational resources—rather than as users of stable systems— in a situation where a multiplicity of representational modes are brought into textual compositions. All the circumstances call for a new goal in textual (and perhaps other) practice: not of critique, but of design (1999, p. 87).

For Kress, design

> takes the results of past production as the resource for new shaping, and for remaking. Design sets aside past agendas, and treats them and their products as resources in setting an agenda of future aims, and of assembling means and resources for implementing that (1999, p. 87).

What Kress adds to the characterization of design is the relationship individuals have with received texts.

Literacy researchers understand that "designing" is useful for understanding processes of communication and meaning making. The New London Group (NLG) (2000), of which Kress is part, argues that design "has become central to workplace innovations, as well as to school reforms for the contemporary world" (p. 19); they believe "design" to be "a sufficiently rich concept on which to found a language curriculum and pedagogy" (p. 20). Because they are concerned with education, the NLG shifts our focus from designing as a process for producing objects to designing as "meaning making" that involves

- "Available Designs," which are the conventions and past texts (as well as past experiences) with which designers are familiar and on which they draw in making new meanings (pp. 20-21).
- "Designing" is the process of "re-presentation and recontextualization" of Available Designs: "Configurations of subjects, social relations, and knowledges are worked upon and transformed" in designing (p. 22).
- "The Redesigned" is what results from an Available Design (or Designs) that has gone through design processes; the redesigned is "never a reinstantiation of one Available Design or even a simple recombination of Available Designs; the Redesigned may be variously creative or reproductive in relation to the resources for meaning-making available in Available Designs" (p. 23).

Anything we produce, mp3 players or brochures, draws on the objects and practices with which audiences are familiar, and also transforms our relations with those objects and practices. The more we and students are alert to these potentials as we research who will use what we make and when, where, and why, the more effectively we can produce transformative designs.

Designing is embedded in the observation of cultural practices. As design anthropologist Tim Plowman (2003) argues, objects—for practical use or communication—engage us precisely because they have their own

> utility as well as their cultural location—the "situatedness" through which designed artifacts recursively derive their meaning and are simultaneously the object of interpretation. In other words, "situatedness" means the multiple ways people consume and integrate designed artifacts into their lives through interaction (use and embodiment) and through their experience create understanding (pp. 30-31).

Plowman reminds us that designed objects are "materialized ideologies," coming out of and fitting back into cultural practices; design's transformations thus open possibilities for changing our environments. Similarly, Charles Kostelnick and Michael Hassett (2003) describe how "conventions"—another term for "Available designs"—work in primarily visual texts:

> By viewing conventions as social constructs underpinned by communities of users, practitioners can more fully understand the rhetorical work conventions do within those social contexts. Identifying variations in the profiles and currencies of conventions can enable designers to make more informed decisions about which ones to deploy, about how to adapt them flexibly to different situations, and about how to blend them in a given document. With a deeper insight into how conventions work, designers can weigh the benefits and liabilities of invoking or flouting certain conventions, and they can aggressively search for new ones that meet the expectations and interpretive frameworks of their readers (p. 229).

In a class that introduces students to design processes, such as I will describe below, it is more than ambitious to hope students will gain such fluency with cultural conventions—but the necessary first steps toward transformative designs lie in helping students learn to think as designers as they work with conventions.

Designing requires attention to the various modalities and materials available for any communication context. The New London Group's (NLG) attention to designing arises, in part, because they see change in the materials of communication. New media, for example, emphasize the visual aspects of texts, such that communicators now often choose not only which words to use but also which photographs, drawings, type, color, or page shape; communicators need to understand how to use the possible spatial arrangements among those various elements. Communicators need also to make decisions about sound and its communicative potentials, because sound is an available resource in digital texts. The NLG thus argues that we need to become as familiar with the visual, gestural (bodily), spatial, and auditory modes of communication as we are with the linguistic (2000, pp. 23-30).

DESIGNING IN TECHNICAL COMMUNICATION

To bring design processes into technical communication classes is to intermingle what might be familiar to technical communication teachers—the rhetorical—with the unfamiliar—the material aspects of texts.

Design is actively rhetorical. Although they do not use the term, descriptions of design from business and industry show designing to require rhetorical savvy, as earlier quotations show: design is a process by which designers engage with and study audiences and the contexts in which audiences move in order that the designers can develop a product—often something whose function is to communicate—for a particular purpose. Practitioners recognize that without respect for the complexities of audience's lives—the contexts within which people move socially, culturally, politically, economically, bodily, and so on—it is impossible to create fitting objects or communications.

How do design and rhetoric differ? Many technical communicators already use rhetorical approaches to help students analyze and produce documents. Approaching technical communication from a design perspective can enrich those approaches because

- **Design approaches attend to the materials of communication.** Because design approaches have developed out of disciplines that produce objects, designers deliberately explore materials they might use. To audience, context, and purpose—the usual rhetorical triad—designers add material and so consider how expected (or unexpected) materials support an audience's use and understanding of a product. Considering different materials can also help designers reconsider initial assumptions about relations among audience, purpose, and context.

- **Design approaches can broaden the scope of research.** Because design has been tied to the development of useful (instead of readable) objects, it tends to foster a more concrete and bodily sense of audience, purpose, and context than rhetorical research often does. Technical communicators do observe communication in workplaces, but the tied-to-bodies practices of design can help technical communicators develop an even richer attention to how, for example, emotions function in learning, or how the size of a document can encourage particular bodily postures. Because design approaches can help designer/communicators give more attention to the material of texts, it can help them better understand the economic aspects of text production, as well as how different materials carry different resonances for different cultures.

- **Design approaches can encourage more active research.** Encouraging technical communication students to attend to the bodily and material aspects of design can help them understand research as including physical observation and action. As they manipulate materials, students are pulled bodily into the design process and so better understand their audiences as real beings; as they observe and interact with people using what they produce, they can more actively attend to the effects of their productions and often listen more closely to feedback.

- **Design approaches can result in more effective communication.** Designers usually approach communication projects without expectations about the final product; their charge is to develop what will work most effectively—and so they experiment. They do not, for example, start by assuming need for a brochure; instead, they study their purpose, context, and audience and consider different possible communication objects for the rhetorical situation. By exploring and testing possibilities, they are more likely to develop what fits.

Designing requires careful and thoughtful observation. The words of the NLG and of Kostelnick and Hassett help us recognize that the workings of any modality, whether we are pondering the register to use in a written instruction set for software or the tone of voice to use in a video about safe use of an industrial drill, depend on the expectations audiences bring to the situation. In designing new communications, we research what audiences know and expect (and what they are dulled to) but this means observing audiences at work.

ASSIGNMENT SEQUENCE: DESIGNING TECHNICAL AND SCIENTIFIC COMMUNICATIONS

Because technical communication students are generally accustomed to producing texts for audiences who do not expect the materials of the communication to have rhetorical function, students can be habituated to *not* seeing the material possibilities in texts they produce. Computer screens or paper can simply seem to be carriers of meaning rather than contributors to an audience's learning. It is helpful then for students to start working with texts where the material of the text does have an obvious and significant rhetorical effect so that students can see the role of such materiality. The assignment sequence that follows thus asks students first to design a book that has interactive elements and that teaches scientific or technical information; students start with a kind of text with which they are somewhat familiar but which they've also probably never produced. They then move to designing a physical learning environment, which is usually more foreign to them and therefore seems to them less constrained by convention, but which also requires more careful consideration of audience and context.

A Note About Using This Assignment Sequence with Majors Outside of Technical Communication and in an Introductory Class

When I have offered this class, it has been taken by technical communication as well as engineering, technology, and business majors. At the end, those from other majors have commented specifically on the usefulness of the design process for their work; in their reflective writings they considered how they would modify what we have done for their discipline's particularities. What perhaps mattered most to me is their determination to approach upcoming work with the openness to multiple and creative solutions that our class encouraged.

I have offered this particular assignment sequence in upper-level classes, but I have also modified the sequence for introductory-level classes: I have asked students in introductory classes to make only the book; at the introductory level, we spend more time talking about how and why to research audiences, for example, and I give them more time to learn how to do such research.

Sequencing and Organizing Instruction

I have described the overall order of the sequence above—book to learning environment (to digital environment)—but within those individual projects is another sequence. For each project, students

1. Reflect on their own learning experiences and then work with others to develop preliminary observations about the characteristics of effective communication for the object they will design.
2. Choose and research a topic to teach.
3. Observe their classmates (their intended audience) using different objects that support the learning the students are to facilitate.
4. Interview their classmates on what they do (or do not) know about their topic in order to identify audience need relative to that topic.
5. Develop a preliminary plan for their communication and test the plan with their audience.
6. Produce a storyboard.
7. Develop a grading rubric for the project.
8. Test the storyboard using the grading rubric.
9. Develop and test a prototype of the project.
10. Develop and test a final version of the project.
10a. Assess the learning that happens. (Note: This step is incorporated starting with the second project; I have found that students understand this step more easily after they have worked through the other steps once.)
11. Reflect on their initial observations about learning as well as on what they learned from this process.

Class Structure

Using design approaches changes the dynamics of the classroom, making it a studio in which students work together to explore communication possibilities. Students tend to work more independently as they get caught up in development, but they also tend to work more collaboratively, seeing the importance of not relying solely on their own observations about audiences. As teacher, I set the framework, but I have to give students room to explore and learn from each other. Given our school's time structure, I teach this class as two 90-minute sections a week in a 14-week semester.

Assessment

As you will see in the assignment descriptions in this chapter's Appendix, throughout the class I ask students to reflect on the design process and their learning, as well as to assess informally the quality of their work and learning. For more formal assessment, students and I develop grading rubrics together, which serve several functions in addition to helping us decide final grades. After they have done enough preliminary development to sense how the object they are building should work, students need to articulate what counts as effective communication of the kind they are designing. I ask them to make their own lists of the qualities they think the final projects should demonstrate, and then we discuss and compile the lists, structuring and categorizing them into a grading rubric on which everyone agrees. In the final stages of project development, students use the rubrics as checklists to ensure they are paying attention to what matters; students also use them for giving feedback to each other. I use them for giving final grades on the projects, compiling my comments with those of others who have also given feedback.

APPENDIX A

PROJECT 1

Designing a book that has interactive elements and that teaches scientific or technical information. [See Figure 2 for and example of a book produced by a student.]

Step 1. Reflecting on your own learning and learning about how others learn

[A handout for an at-home assignment.]

Reflecting on Your Own Learning

Part 1—Call to mind a time you really learned something, and learned in a way that was both pleasurable and effective. Describe in detail both what you learned and how you learned it. Were other people involved in your learning, or were you on

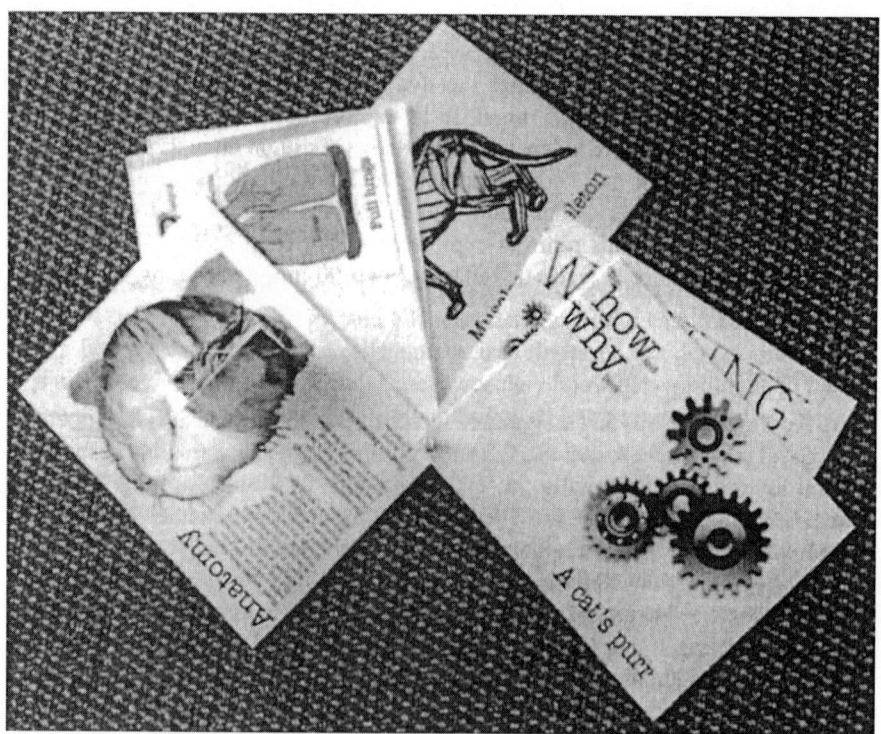

Figure 2. A book made by Jana Jones to teach how and why cats purr. Through a series of lift-up flaps and transparencies, this book explains the physiology of a cat's purr. Reprinted with permission of Jana Jones.

your own? How did you feel while you were learning, and what contributed to your feeling that way? How did (did?) the physical situation around you contribute to your learning? How did what you learned matter to you—and why? Please type 500-750 words that respond to these questions—and at the end of your writing make some general speculations about conditions that encourage your learning.

Part 2—Call to mind books or learning environments (classrooms, museums, libraries, movies, educational games, your kitchen) that have supported you in learning. List as many features as you can of the books or learning environments that encouraged your learning. Also describe how you think the context in which you came to the book or environment helped your learning: were you already curious about the topic, and/or did the pop-ups or textures or writing in the book catch your attention and get you interested, and/or did hands-on features or the order of presented information in the museum draw you in, or . . . ? Please type 500-750 words that respond to these questions—and at the end speculate about conditions that encourage your learning.

Learning about the learning of others

[A handout for an in-class activity.]

Talk with others in class about their own learning memories. Together, identify what is in common among your memories and experiences. After you've talked with 4 to 5 others, work with a partner to list your observations about what makes learning engaging, based on your conversations. This list will be a working reference for you as you develop the upcoming project, so that you can test and revise your observations.

Step 2: Choose and Research a Topic

[A handout describing out-of-class work.]

Part 1—Choosing a Topic

Why I Ask You To Do This

When you hear "book" you probably think of something hardbound that fits in your hands and contains pages of print: a novel, perhaps, or a book of theory. But there are also comic books, children's books, auto repair books, travel guides, cookbooks, pop-up books, artists books . . . objects bound in all sorts of ways with lots of different things on their pages.

We will use "book" in this class to mean pages (not necessarily made of paper) bound together in some way. I want you to explore how you can use a book-object you make to teach creatively a scientific or technical topic. I want you to think of the book you make as an experience for your intended audience rather than only as a set of words. Books are relatively easy to make, requiring no specialized technical facility, so you can comfortably and creatively explore the possibilities of this object for teaching.

What To Do

Pick a topic from the list below that you yourself do not understand. In developing your book, you will have to learn enough about the topic to explain it to others—which means you need to attend to how you learn and hence to how others might learn as well; this also means that you are more likely to have reasonable expectations about what your audience already knows.

For this project, your audience is a non-specialist adult audience; the class can then act as your audience for testing and helping you revise.

You can design your book using any materials to which you have access, and it can be in any format; the only requirements are that your book must be a series of bound pages (remember, not necessarily paper) and that you can justify your design decisions in terms of your purpose, context, and audience.

The book needs to be a stand-alone: others have to be able to read and understand it without you being near.

NOTE: You need to make three copies. This will help you understand how the material construction of the book supports it becoming a mass produced object (or a singular gift for one other person) as well as the time and material constraints of designing communications. After you graduate, you will be expected to produce communications that are inexpensive and easy to produce—but that also engage audiences effectively. By having freedom with materials and construction now, you will sense how to work with the constraints that can get in the way of (or encourage) the most effective communications.

Possible Technical/Scientific Topics to Teach

The water cycle • digital data storage • the combustion engine • human digestion • digital paper & ink • precision bombing • natural selection • lake effect snow • steroids • vacuum cleaners • the Krebs cycle • escape velocity • wireless telephones • viral reproduction • dirty bombs • night-vision goggles • neurochips • how the magnetic axis of the earth flips over time • prions

[The following is a handout for further outside-of-class work.]

Researching Your Topic

After you have a topic, learn as much as you can about it. Consult at least 5 sources, one of which must be a human who is expert on some aspect of the topic. On our campus are people who know about the topics I listed: you do not need to interview this person extensively; instead, contact this person to see if he or she can confirm that you understand what you are teaching so that you teach it accurately and can suggest other perspectives or approaches than you have been considering. (If you need help in identifying and contacting such a resource, I can help you.)

After you've carried out your research, type a 500-1000 word description of what you will teach, and why. Describe the process or concepts you are teaching. Describe why understanding the process or concepts should matter to your audience. At the end, list your sources in either MLA or APA format.

Step 3: Informally Observing How Your Audiences Uses Book

[The following describes an in-class activity.]

I bring into class a large assortment of technical and scientific books for lay audiences (including children's and young adult books, because these often make use of the material possibilities of books); the books cover topics from car repair to the life of eels. I ask students to look through the books to find ones they like and ones they don't. They work in pairs, developing lists of what they like and don't like about the books, as well as of what they think is effective in encouraging learning and what not.

The class then divides in two: one half is observers, the other half the audience to be observed. Each observer picks one of the books that she or he considers to be effective. The observer then watches an audience member use the book: the audience member is to speak-aloud what he or she is experiencing and thinking while moving through the book; the observer makes notes. I ask observers to look for what engages a learner in the book and what doesn't, what in the book encourages a learner to feel smart and capable as well as what is confusing or dense, and how learners move through the books. After 15-20 minutes, the audience members become observers and the former observers become audience members.

After everyone has been both observer and audience, students take 5-10 minutes to work in pairs to draw up speculative conclusions from their observations, based on the characteristics I asked them to observe.

Finally, I ask students to reflect on this activity as an act of purposeful observation: How could we have structured this class project to help them see more usefully how others use books for learning? They then draw up descriptions of how they would approach such observation in the future.

Step 4: Interviewing to Learn What an Audience
Does or Does Not Know . . .

[A handout setting up an in-class activity.]

Now that you have researched your topic and speculated about what your audience does and does not know, and you have watched how others use books in learning, it is time to consider how your particular book will work. Interview at least 6 other people in class, to learn the following:

- What do members of your audience know about the topic you are teaching?
- What are members of your audience curious about, regarding your topic? What might encourage them to be curious?
- What ways of contextualizing your topic will help your audience understand that the topic matters to them?
- What kinds of interactions might best help them learn about the topic?

Before you start interviewing, take 10 minutes, working with a partner, to develop approaches to interviewing that you think will help you learn useful responses to these questions. Should you describe your topic in detail and then ask the questions above? Should you simply ask people you interview what they know about your topic? While you interview, take notes, and record any suggestions your interviewees make about teaching strategies for your book; also note when interviewees show enthusiasm or boredom.

After interviewing, take several minutes to write a quick summary of your observations from the interview (I won't collect this, but be sure you can read it). In your writing, consider the following: Do you need to modify or change what you were planning on teaching, either by going deeper into the topic or by taking a more introductory approach? What about your topic interested your audience? What aspects of your topic will you have to make more compelling to your audience? How can you show them that they should care about your topic?

After you finish your reflections on what you learned, meet for a few minutes with the partner who helped you develop your approach to the interviews. Write up for me (and this I will collect) a quick description both of what you think worked in your interviewing approach, and what you will do differently next time.

Step 5: A Preliminary plan for your book.

[A handout describing an out-of-class assignment.]

Why I Ask You To Do This

The questions I ask you to address below should help you think concretely about the book you are designing.

What To Do

Bring to class typed pages (750-1000 words) in which you respond to the following:

1. What is the scientific or technical topic you'll teach in your book?

2. Look back over the writing you did at the beginning of class, about how learning can be effectively structured. Based on your reflections, and on the discussions and observations you've made in class, what kind of experience(s) do you want your book to provide your audience? (That is, as they move through your book, what should they be thinking or feeling? What sorts of expressions should be on their faces? How should they be moving?) How do the experiences you describe support the particular learning you want your book to perform?

3. Sketch out (in words) at least 3 different approaches you could take with your book. Sketch out 3 very different approaches to writing what could be in the book and to showing visually and/or interactively what the book could be. The following questions will help:

What concrete design directions are suggested to you by the experiences you want your audience to have? Do you want your book to teach straightforwardly, addressing the audience directly as they are, or should the book ask the audience to step into an imaginary scenario? Should the book be written in the third or first person? What kinds of materials will support the experiences and learning you seek? Should the book be big or small? Should it use color or not? Should the book be rectangular or . . . ? What should the ratio of words to photographs or drawings be? What ought to be interactive about the book? Consider as many different directions as possible, trying to make each one work. Based on the experience(s) you want your audience to have, what are several different ways to order or chunk the information you are teaching?

4. How do the 3 different approaches you are considering support you in providing the experience you plan for your audience, help them learn the topic you've chosen, and provide a memorable learning experience?

Testing The Plans

In class, describe your three different approaches to the book to someone else. After describing, ask which approach seems most likely to support an audience member in learning. What is it about the approach that seems to work? Would your book be more effective if it combined some of the approaches? Talk with at least 4 other people.

Step 6: Making a Storyboard for the Book

[A handout describing an out-of-class assignment.]

Why I Ask You To Do This

It's time to make a decision about the approaches you have been considering. Making a storyboard for your book will help you work out both practical and teaching issues before you are in the middle of the process of building and do not want to start over because something isn't working properly . . .

What To Do

[NOTE: The descriptions below could imply that you need to build your storyboards in the order of description. Build your storyboards in whatever order makes sense to you.] Your storyboards should contain the following:

- FIRST, an introduction to the storyboard in which you tell the name and purpose of your book, describe the book's final size as well as the materials you will use, and describe how you will bind the book.
- SECOND, as many pages as you are now planning your book to have (and be sure to count the cover of your book as one of its pages). Each page should have on it:
  – A rough sketch of how the page will look. (And I emphasize "rough" here: color is not necessary, nor are highly finished pages. All you need is enough

information so someone else can give you feedback; if you make these sketches too polished and detailed, you'll become too wedded to your work and will find it harder to revise the pages in response to feedback.)

   – A description of any kind of activity an audience can take on or with the page.

   – A short justification for why the page is designed as it is and has on it what it does. How will what is on this page help your audience learn?

   – A description of the reflection you will ask your audience to undertake to be sure they learned what you need them to learn.

- FINALLY, include a page on which you justify your major design decisions in terms of your purpose (the experience and learning you want to provide for your audience).

Step 7: Developing a Grading Rubric

[A handout describing an in-class activity.]

Using your storyboards to help you imagine the finished book and how people will use it, list the qualities that you think the book—in order to be effective—ought to have. Compare your list with a partner, and merge your lists into one.

We merge their lists to make a single rubric, such as the one in (such as the one in Figure 3).

Step 8: Testing the Storyboard Using the Rubric

[A handout giving instructions for an in-class activity.]

Give your storyboard to someone else in class; after that person has looked through it, ask the reader to explain back to you what you are trying to achieve and how. Are there any questions you need to answer about the storyboard or clarifications you need to make? You also will look at someone else's storyboard. As you look over it, imagine as concretely as you can how the book looks and feels. Is there any part you can't visualize because you don't have enough information? Are there parts whose ordering you don't understand? Come up with as many questions as you can about how the book functions, so that you can be as helpful as possible to the storyboard maker. Get answers to your questions from that person.

Once you are sure you've understood the storyboard enough to visualize it concretely, fill out a rubric for it. As you fill out the rubric, be sure to make comments that suggest changes. For example, do not write, "I don't like this" or "This is cool" or "I don't get it"; instead, write "I don't think [name the aspect of the book to which you are responding] will appeal to your audience because . . ." or "I think [name the aspect of the book] will be particularly effective because . . ." or "I don't understand why the book does [x] because I thought your purpose was [y], and here's why I think [x] doesn't support [y]." As you write your comments, please use a supportive tone of voice; the purpose of doing this work is to help each other make the coolest, most effective books.

Your name:

Name of person who made the book you are reviewing:

Grading rubric for the book

quality	does this perfectly/ beautifully	does this better than expected	does this	does this, sort of	doesn't do this yet	N/A	comments
The book							
is visually appealing							
is effectively interactive for its audience and purpose (as our reading in the book *Experience Design* defines it, "interactivity is a process of continual action and reaction between two parties" and involves "feedback, control, creativity, adaptivity, productivity, communications. . . ." (142))							
is organized to support the learning that the audience is supposed to do by using the book							
is engaging for its intended audience.							
teaches what it sets out to teach.							
is usable & sturdy							
is constructed with a quality appropriate for its audience.							
The information in the book							
has an appropriate tone for its audience.							
is formatted appropriately for the audience.							
is appropriately designed for what the book is trying to teach							
The writing in the book is polished and grammatical							
The visuals in the book (photographs, illustrations, etc.) are of appropriate quality for the audience & context.							
Overall, the book...							
shows the effort its maker put into it.							
is appropriately creative							
stimulate's audience thinking (by encouraging them toward new thinking on a topic or looking at something familiar in a new way).							
is memorable							

Please indicate here any specific actions you think the book's maker can take in order to improve the book in the areas where the book was not perfect.

Figure 3. Sample rubric.

[In class, after everyone has received a completed rubric, students write a reflection on what works in their proposed books and describe what changes they will make. As a class, we discuss how well the rubrics worked and make any modifications that seem necessary. I also use the rubrics to give feedback to each of the storyboards.]

Step 9: Developing and Testing a Draft of the Book . . .

[A handout for an out-of-class assignment.]

Why I Ask You To Do This

I want you to make a draft so you can get feedback on it and have time to revise it—so that the final book is as complete, polished, and effective as possible.

What To Do (And Not . . .)

DO NOT use the final materials you were planning to use, and do not get as elaborate as you were planning with binding or pop-ups or . . . Use scrap paper and staples and pencils or crayons or . . . Build a mock-up of your book that is complete enough to give others a solid sense of what you are planning but that is not so complete that you will be loathe to make changes to it.

[In class, students use the grading rubric to give feedback on the drafts. After receiving three completed rubrics, students write a reflection on what works in their drafts and describe what changes they will make for the final.]

Step 10: Developing and Testing a Final Draft of the Book

Students bring final drafts of their books to class. (They have had to make three copies of it.) We invite people from outside the class to come and look at the books, and to fill out rubrics. Students in the class also fill out rubrics for each other's books.

Step 11: Reflecting on the Process . . .

[A handout for an out-of-class assignment.]
Please write a 750-1000 word response to the following questions:

- If you could take your book through one more revision, what would you change?
- Look back to the initial observations you made about situations that support effective learning. How well did these observations help you design the book? Which observations were most useful? From building and testing your book, what other conditions do you see that support effective learning?
- What did you learn about the time that goes into producing a book? What would you need to change about your book if it were to be mass produced?
- How did the materials you used for your book support the learning of your audience?

- What aspects of the design process, as we set it up and experienced it in this class, helped you learn effectively? What aspects of the process helped you most in developing a creative, effective book? What ought you to do differently next time?
- How do you think you can learn more about what audiences know and expect about a topic for which you are building communication?
- What would you do differently the next time you had to design a book for teaching technical or scientific information?

PROJECT 2

Designing a learning environment for teaching scientific or technical concepts

This project follows the same steps as the book project, except that students extend their research and—in order that they develop more confident and informed independence—they take more responsibility in shaping the research; they also develop strategies for assessing that audiences do indeed learn what the environments are intended to teach. For those reasons, I have only included below descriptions of steps that differ from the book project. Please see Figures 4–5 for samples of these learning environments.

Step 1: Reflecting on Learning Processes

[A handout for an out-of-class assignment.]

ANALYZING A LEARNING ENVIRONMENT

Why I Ask You To Do This

You are going to build a learning environment, some sort of object that engages as many of our senses as possible. You will get ideas about how to make such an environment as effective as possible by analyzing what works (or not) about a learning environment you've already experienced.

What To Do

Choose a learning environment you've experienced: a class or lab, an exhibit in a museum, a chemistry set you once had, a toy that helped you learn about shapes, color, or the alphabet . . . any learning experience that involved being in a discrete space and/or using discrete objects you could hold or smell or taste or . . . Type 1-2 pages in which you name and describe the environment, and then analyze it using the following questions:

- Was the environment effective for you? Why?
- Was the environment memorable for you, pleasantly or not? Why?
- How was your body placed in the environment or relative to the environment? What bodily actions did you need to take in order to learn from the environment? How did your bodily actions contribute to your learning?

Figure 4. A kit made by Jana Jones to teach about Multiple Sclerosis.
Reprinted with permission of Jana Jones.

- What were the experiences you were designed to have in this environment?
- What would you change about the environment to have made it a more effective
  environment?

Step 2: Choosing and Researching a Topic

[A handout for an out-of-class assignment.]

DESIGNING A LEARNING ENVIRONMENT

Why I Ask You To Do This

When you hear "environment" you might think that I am asking you to fill some
large physical space with objects that an audience touches or holds or shakes. This is
a possibility for satisfying this project's requirements (if you can find a space we can
get to easily). But I'm using "environment" because I want you to think of bodies
in space, to think of the possibilities of engaging other senses besides sight in

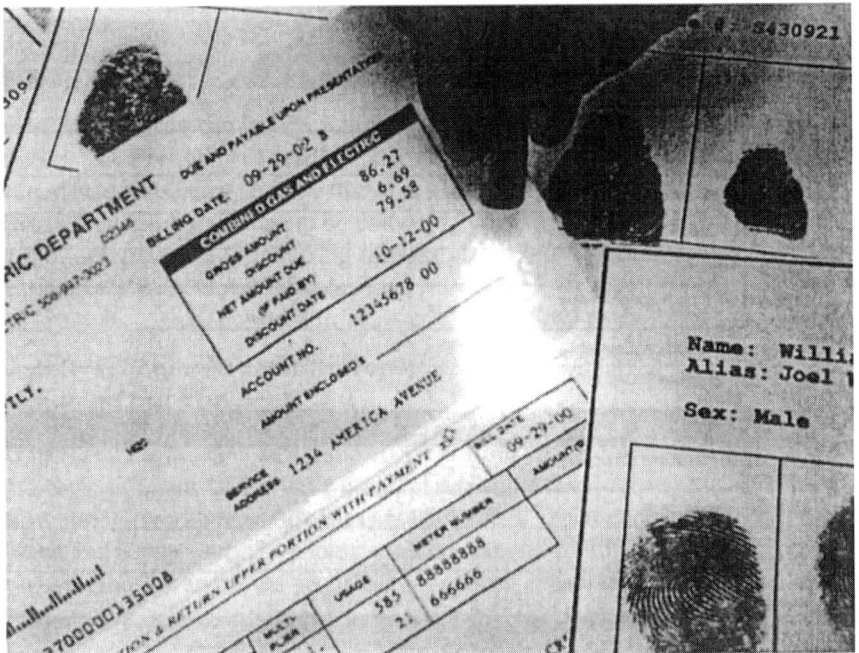

Figure 5. A kit made by Katie Cleary to teach about crime solution technologies. The above materials were contained in a boxed simulation in which students learned about a crime and then were given documents and tools—including a device for seeing fingerprints through certain lighting conditions—that enabled them to solve the crime. Reprinted with permission of Katie Cleary.

learning, and to think about how movement and sound can encourage effective and engaging learning.

We'll be looking at examples of environments in class, but—as you think about this project—let your thoughts wander to any object or place or piece of clothing or toy or museum display you've experienced that captured your imagination and made you forget time. Then think about how you might reproduce that experience to teach scientific or technical concepts.

What To Do

Pick a topic from the list below that you yourself do not understand. In the process of developing this environment, you will have to learn enough about the topic to explain it to others—which means you have the chance to attend to how you learn and hence to how others might learn as well; this also means that you are more likely to have reasonable expectations about what your audience already knows.

For this project, your audience is a non-specialist adult audience, so that we in class can act as an audience for testing and helping you revise this project.

Your name: _____ Name of person(s) who made the environment you are reviewing: _____

ASSESSMENT RUBRIC FOR THE LEARNING ENVIRONMENT

The environment...	5=perfect	comments
introduces the audience to the topic (brings them into the topic) in an engaging way.	5 4 3 2 1	
keeps its intended audience drawn in — they don't walk away after 2 seconds.	5 4 3 2 1	
teaches what it sets out to teach.	5 4 3 2 1	
is designed to support its teaching purposes.	5 4 3 2 1	
is effectively interactive.	5 4 3 2 1	
can be used without outside explanation or help.	5 4 3 2 1	
is well constructed, usable, and sturdy.	5 4 3 2 1	
is memorable / has a large "WOW!" (or "GROOVY!") factor.	5 4 3 2 1	

The scientific/technical material...

is appropriately (for audience and purpose) informative.	5 4 3 2 1	
is effectively communicated.	5 4 3 2 1	
is effectively integrated with the design of the environment.	5 4 3 2 1	
has a tone appropriate for the audience and purpose of the project.	5 4 3 2 1	

The design of the environment...

helps the audience create knowledge from the information presented — or look at the information in a new way.	5 4 3 2 1	
helps the audience retain the knowledge they learned from the environment by asking them to reflect on or apply what they learned.	5 4 3 2 1	
addresses the senses appropriately & engagingly for what is being taught.	5 4 3 2 1	
shows the effort its maker put into it.	5 4 3 2 1	
is appropriately creative.	5 4 3 2 1	

Any writing in the environment...

is free of mechanical/grammatical errors.	5 4 3 2 1	
helps the audience understand how different pieces of writing are connected.	5 4 3 2 1	

GENERAL COMMENTS on what is most successful about this project and what could be strengthened:

Figure 6. Sample Rubric for the Learning Environment.

You can design your environment using any materials to which you have access, and it can be in any format and size; the only requirements are that your environment must address as many human senses as possible and must pull in your audience's bodies, and that you can justify your design decisions in terms of your purpose and audience. The environment needs to be a stand-alone. That is, others need to be able to experience and understand it without you being anywhere near.

Because environments tend to be awkward things to include in a portfolio or bring to a job interview, we'll be talking in class about how to document your environments, and we'll document them in class.

Possible Technical/Scientific Topics to Teach in an Environment

[**NOTE:** You should choose a topic different from your book; if you want to work with a topic different from what's listed, please come talk.]

How springs work • tectonic plates • human senses • solar panels • why does music make us feel happy or sad? • photosynthesis • nanotechnologies • 3D movies • how the Mars lander transmitted its images • gender selection and microsort sperm separation • light bulbs • the senses of insects • how the magnetic axis of the earth flips over time • microbanking • retinal implants • memory in the human brain.

Step 5: Developing an Initial Plan for the Learning Environment

[A handout for an out-of-class assignment.]

Why I Ask You To Do This

This will be your first production on the way to building the learning environment, a plan to which you can get feedback to help you make the environment more effective, memorable, pleasurable…

What To Do

1. What is the scientific or technical topic of your environment? What specifically do you want your audience to learn?

2. Look back over the writing you did at the beginning of class and at your reflections from designing the book, about how learning can be effectively structured. Based on your reflections, and on discussions and observations from class, what kind of experience(s) do you want your environment to provide your audience? (That is, as they are moving through your environment, what should they be thinking or feeling? What sorts of expressions should be on their faces? How should they be moving? What should their experience of the space of the project be? That is, should the environment feel close and comforting or vast and overwhelming?) How do the experiences you describe support the particular learning you want your book to perform?

3. Sketch out (in words) at least 3 different environments that could support your audience in learning what they need to learn. Sketch out 3 very different approaches to the environment, showing visually and/or interactively what the environment could be. The following questions ought to help you think toward this: What concrete design directions are suggested to you by the experiences you want your audience to have? Do you want your environment to teach straightforwardly, addressing the audience directly as they are, or should the audience step into an imaginary scenario? What sorts of materials will support the experiences and learning you seek? Should the environment be completely self-contained (like a suitcase) that the audience opens up or should it be something the audience steps into? Should it be colorful or sedate? What will you convey in writing and what through interactions or illustrations? Consider as many different directions as possible, trying to make each one work.

Based on the experience(s) you want your audience to have, what are several different ways to order or chunk the information you teach?

4. How do different approaches you are considering support you in providing the experience you plan for your audience, help them learn the topic you've chosen, and provide a memorable learning experience?

Step 6: Developing a Storyboard for the Environment

[A handout for an out-of-class assignment.]

Why I Ask You To Do This

It's time to make a decision about the different approaches you've been considering. Making a storyboard for your environment will help you work out both practical and teaching issues before you are in the middle of the process of building and do not want to start over because something isn't working properly . . .

What To Do

[**NOTE**: The descriptions below could imply that you need to build your storyboards in the order of description. Build your storyboards in whatever order makes sense to you.] Your storyboards should contain the following:

- FIRST, an introduction to the storyboard in which you tell the name and purpose of your environment and quickly describe how the environment will look and of what it will be made.
- SECOND, make a "sketch" of each conceptual or physical step an audience member will undertake as they move through your environment. Do this by taking photographs of friends in each of the postures they need to undertake as they move through your environment; that is, photograph the friends as though they were holding or interacting with the environment, but without the environment—this requires considerable concrete imagining on your part, and so is a

highly useful approach to thinking through exactly what the environment needs to do. Here's how to present the photographs:
– Mount each photograph on its own page.
– Under the photograph, write a description of what the person is doing with the environment, what they should be experiencing/thinking/learning.
– Under the photograph, write a short description of why you have arranged the environment such that someone from your audience has to move through it in the order you show. (Or, if you've designed a non-linear environment, describe why the environment is non-linear and how the audience will be able to learn from the non-linearity)
– Provide a description of the kinds of reflection you will ask your audience to undertake to insure they learned what you want them to learn.
• FINALLY, include a page on which you justify your design decisions in terms of your purpose (the experience and learning you want to provide for your audience).

Step 10a: Assessing the Learning that Happens in the Environment

[A handout for an out-of-class assignment.]

Why I Ask You To Do This

You need to know whether the environment (or any learning materials you develop) is effective, teaching what it should. This assignment starts you toward one process for determining whether people learn.

What To Do

Type 1-2 pages in which you do the following:
1. You cannot assess whether people learned something unless you know what it is you are hoping they learn—so describe in detail the exact information you want your audience to learn. List facts, figures, concepts, and/or processes that your audience should know and understand as a result of working with your learning environment.
2. Describe in detail the kind of emotional connection or attitude you want your audience to have with or about what you described in #1. Because we are trying, with these environments, to help others learn in more engaged ways than often happens, you need to be able to specify the kinds of engagement you want your audience to have.
3. Brainstorm several ways you could test whether you audience has learned what you described in #1, with the attitude you described in #2. Think of paper tests, or of asking people to construct or describe something, or to sketch a process, or to change a situation. Think of any action (which includes writing) that your audience could take that would demonstrate that they understood what you described

in #1, and that they had the attitude of #2. [Do not just list "paper test, construct something. . . ." Instead, list the questions you would ask on the paper test; describe the conditions under which the audience would construct something without your help, etc.]

4. Consider when it would be best for your audience to be assessed on their learning. Should it happen immediately after they've used the environment, or a day or two or seven later? Why?

In our next class we will discuss and develop a group approach to assessment.

Step 8: Developing a Rubric for the Environment

Figure 6 shows one possible rubric for the environment.

Step 9: Developing and Testing a Prototype of the Learning Environment

Why I Ask You To Do This

By building a "quick" mockup of your environment, you'll be able to get feedback and so be able to revise your plans to make the most effective and memorable learning experience.

What To Do

I cannot give you hard and fast instructions for this because each of you is constructing such different environments. What is most important is to construct the quickest version (not necessarily to scale) of your environment that will give others enough sense of the salient features so they can give you useful feedback. Use any materials available to you for this—and please come talk if you need help.

REFERENCES

Butkus, M. A., & Kelley, M. B. (2004). Approaches for integrating professional practice issues into undergraduate environmental engineering design projects. *Journal of Professional Issues in Engineering Education and Practice, July,* 166-172.

Davis, M., Hawley, P., McMullan, B., & Spilka, G. (1997). *Design as a catalyst for learning.* Alexandria, VA: Association for Supervision and Curriculum Development.

Kress, G. (1999). *"'English' at the crossroads." Passions, pedagogies, and twenty-first century technologies* (pp. 66-88). G. Hawisher & C. L. Selfe (Eds.). Logan, UT: Utah State University Press.

Kress, G. (2003). *Literacy in the New Media Age.* London: Routledge.

Kostelnick, C., & Hassett, M. (2003). *Shaping information: The rhetoric of visual conventions.* Carbondale, IL: Southern Illinois University Press.

The New London Group. (2000). A pedagogy of multiliteracies: Designing social futures. In B. Cope & M. Kalantzis (Eds.), *Multiliteracies: Literacy learning and the design of social futures* (pp. 3-37). London: Routledge.

Partnership for 21st Century Skills. (2003a). Learning for the 21st century. Available online at: www.21stcenturyskills.org.

Partnership for 21st Century Skills. (2003b). The road to 21st century. Available online at: www.21stcenturyskills.org.

Plowman, T. (2003). Ethnography and Critical Design Practice. In B. Laurel (Ed.), *Design research: Methods and perspectives* (pp. 30-38). Cambridge, MA: MIT Press.

"Project Missions." (no date.) Feral Robotic Dogs. [Website.] http://xdesign.ucsd.edu/feralrobots/projectindex.html.

CHAPTER 5

Copyright Law and a Fair Use Pedagogy: Teaching and Learning Strategies for Technical Communication Courses

Michael R. Moore

> **OUTCOME:** Students will be able to identify contemporary contexts and issues in copyright and fair use, and articulate those contexts and issues for multiple professional audiences.

"Intellectual Property" refers to concepts such as trademarks, trade secrets, patents, and copyright. As defined in The Copyright Act of 1976, intellectual property encompasses literary, dramatic, and musical works; pictorial, graphic and sculptural works; audio/visual works; sound recordings and architectural works ("Copyright Act," §107, 1976). Recent legislation has introduced major changes in U.S. copyright law to address digitally networked environments; The Digital Millennium Copyright Act ("DMCA"), for example, makes amendments to the Copyright Act of 1976, including amendments designed to protect against circumvention of copyright-protection systems.

Many people, however, especially those who use internet resources, argue that such legislation is overly restrictive and that it should also protect the fair-use doctrine on the internet and in other multimedia contexts.

Additional unresolved issues associated with intellectual property—the representation of students as criminals, economic overdetermination, and the effects of globalization on communication—also raise new exigencies that should be of interest to teachers and students in introductory technical-communication courses. Because copyright and fair-use issues affect all students and all educators, this chapter attempts to provide some resources and strategies for faculty who teach such courses.

WHAT BUSINESS, INDUSTRY, AND LEGISLATIVE ACTION TELL US

Historically, business and industry sectors in the United States and many Western countries concerned with intellectual property as a valued commodity have focused on intellectual property issues primarily in terms of narrower, more restrictive copyright laws. The Recording Industry Association of America (RIAA) in particular has developed antipiracy campaigns and has filed suit against hundreds of individual file sharers (Webb, 2003, "RIAA vs. the People"). Similarly, the Walt Disney Company has a decades-old tradition of aggressively protecting its intellectual property. Some critics argue that it does so to the detriment of public-domain materials, free-speech activities, and protected parody uses under fair-use guidelines (Bettig, 1996; Lessig, 2000; Levin, 2003; Vaidhyanathan, 2003).

These cases help us recognize **a fundamental link between economic interests and intellectual property concerns in the U.S. marketplace**—one that is, at some level, self-motivated. As the Creative Incentive Coalition (CIC), notes for example,

> Copyright-based industries . . . make a large and growing contribution to America's economic welfare and global competitiveness. Accounting for nearly 4% of the nation's Gross Domestic Product, these industries employ more than 3 million U.S. workers, and are creating new jobs at a rate more than triple the employment growth rate in the economy as a whole" (Kay, 1996, "Comments on Hearings").[1]

In light of this economic context, copyright-based industries have argued for stronger, less flexible copyright laws, based, notably, on global competitiveness.

If, however, many of the major intellectual property cases pursued by business and industry have been characterized by a conservative and protectionist approach to copyright, they have also served to increase the public awareness of the complexity that attends most intellectual property issues. Kembrew McLeod (2001), in *Owning Culture: Authorship, Ownership, and Intellectual Property Law*, for example, notes that, "the fabric of social life in most Western countries—and, increasingly, the world—is becoming more deeply immersed in the domain of intellectual property law" (p. 1).

---

[1] This information comes from the CIC, a "broad-based copyright industry coalition that works to improve public understanding of the importance of copyright in the digital age." The occasion for these remarks was the Federal Trade Commission's Hearings on Global Competition and Innovation in 1995. It's interesting to note the CIC's membership at the time of the FTC remarks: Association of American Publishers; Association of Independent Television Stations; Business Software Alliance; Cox Enterprises; General Instrument Corporation; Information Industry Association; Information Technology Industry Council; Interactive Digital Software Association; International Business Machines Corporation; Magazine Publishers of America; McGraw-Hill, Inc.; Microsoft Corporation; Motion Picture Association of America, Inc.; National Cable Television Association; National Music Publishers' Association; Newspaper Association of America; Recording Industry Association of America; Software Publishers Association; Time Warner, Inc.; The Times Mirror Company; Turner Broadcasting System, Inc.; West Publishing Company; Viacom, Inc.

result of online file-sharing (video, music, text) controversies or as an issue raised within the context of courses in which they are asked to use quoted materials and images in projects and assignments. As Anthony Bates (1999) notes in *Managing Technological Change: Strategies for College and University Leaders*, "nowhere is there more confusion, misinformation, and paranoia than in discussions of intellectual property and copyright surrounding the development and use of digital materials" (p. 107) than on college campuses.

Fair Use

Many students and teachers in technical-communication courses have also had to confront the doctrine of Fair Use, Section 107 of Copyright Law—a provision which allows for some limited uses of copyrighted materials within academic contexts depending on four factors (Section 107. "Limitations").

With the emergence of digital contexts and materials, however, **differing interpretations of fair use have been circulated in online resources, arguments, and materials that appear both in print and online** (Benedict.com; CNI; Copyright.com; Harper "Crash Course"). Such sources are authored by a wide range of people and institutions and promote distinctly different readings of Section 107 of Copyright Law in general.

Within educational contexts, these interpretations of fair use are supplemented with U.S. Copyright Office guidelines for "Reproductions of Copyrighted Works by Educators and Librarians" (Circular 21, 1998). Recent legislative attempts to update or modernize both copyright law and educational guidelines have resulted in the Digitial Millennium Copyright Act (DMCA, 1998), the related Copyright Office Study on Distance Education (Distance Ed, 1999), and the "Technology, Education and Copyright Harmonization Act" (TEACH Act, 2002).

The guidelines and legislation associated with copyright law remain extremely complex. As a result, an attorney representing the Recording Industry Association of America (RIAA), a teacher of first-year rhetoric and composition exploring the social construction of legal discourse, a technical communication scholar, and a professor of computer science researching the reverse engineering of software may well have dramatically different understandings of copyright and fair use, their possibilities, their constraints, and their meanings.

The TEACH Act

The TEACH Act, in particular, provides the most recent curricular and legal guidance about copyright, and it has implications for both face-to-face and online classroom activities. For example, Kenneth Crews, in his "New Copyright Law for Distance Education: The Meaning and Importance of the TEACH Act" (2002) summarizes fair-use history and legislation, noting that "In the context of traditional, face-to-face teaching, educators long have debated the application of 'fair use' to making copies, and the Copyright Act since 1976 has included a relatively simple and broad provision allowing 'performances' and 'displays' in the face-to-face classroom setting" (p. 2).

Crews continues,

> *The rules for distance education, however, are significantly different.* Both the meaning of fair use and the details of the specific statute become much more rigorous when the materials are uploaded to websites, transmitted anywhere in the world, and are easily downloaded, altered, or further transmitted by students and other users—all posing possible threats to the interests of copyright owners (p. 2; emphasis added).

Crews's analysis points to how new intellectual-property guidelines might affect online learning spaces in the coming years. According to Crews, the TEACH Act is

> . . . built around a vision that distance education should occur in discrete installments, each within a confined span of time, and with all elements integrated into a cohesive lecture-like package. In other words, much of the law is built around permitting uses of copyrighted works in the context of "mediated instructional activities" that are akin in many respects to the conduct of traditional classroom sessions (p. 3).

These legislative efforts serve, implicitly or explicitly, to prescribe the overall design of online learning environments. Teachers, within such a system, for example, are assigned much of the responsibility for protecting intellectual property within academic institutions—a tendency which, in turn, contributes to an increasing centralization of power. The potency of the trend is magnified, moreover, by the many distance-education initiatives that have been undertaken on campuses. As a result of these efforts, librarians and campus information-technology officers have also become increasingly responsible for intellectual-property protection under the TEACH Act, contributing to additional centralization of power in connection to intellectual property.

How are students' classroom and learning activities constrained within these same legislative contexts and discourses? Crews notes that under the TEACH Act, "the law anticipates that students will access each 'session' within a prescribed time period and will not necessarily be able to store the materials or review them later in the academic term" (p. 3). **It is difficult to imagine how students will deal with such constraints, when they are enrolled in composition classes that encourage, promote, and require iterative reading and design, the exchange of annotated files and papers, collaborative writing, peer editing, or the collective authoring of multimedia.**[2]

---

[2] This is not to suggest that the TEACH Act is without merit or that it does not address many copyright issues that teachers currently face. To the contrary, in her 2001 remarks to the Senate Committee on the Judiciary, Marybeth Peters, head of the U.S. Copyright Office, noted that, "technological change had made it possible for educators to reach a vastly broader student population with a richer variety of course materials than was ever possible before the advent of the Internet," and that teachers often use technology in the service of a "rich pedagogical experience" (Peters, 2001, "Statement").

Perspectives on Ethics and Ownership

The contested landscape of intellectual-property issues has extended into the academic realm in other areas as well, especially in discussions of ownership. For example, in *Intellectual Property Law for Engineers and Scientists*, Howard Rockman (2004) discusses in great detail the history, ethics, and contemporary contexts for professionals in engineering, science, and technology-related fields, but ultimately concludes that the most important aspect is to learn to *protect* one's intellectual property ("Top Ten List of Intellectual Protection"). Protecting one's property is certainly an increasingly important function in a culture—and in an ethics—based primarily on ownership. However, scholars like Paul Butler have identified alternative views. In his 2004 *Stanford Law Review* article, "Much Respect: Toward a Hip-Hop Theory of Punishment," Butler offers an alternative theory of intellectual property similar to those innovative alternatives acknowledged by Copyleft scholars in multiple disciplines.

> For some time the debate about why people should be punished has been old school: Each one of the four theories of punishment—retribution, deterrence, incapacitation, and rehabilitation—has acceded to prominence, and then lost its luster. Hip-hop offers a fresh approach. It first seems to embrace retribution. 'The unwritten law in rap,' according to Jay-Z, is that 'if you shoot my dog, I'ma kill yo' cat . . . know dat/for every action there's a reaction' . . . Hip Hop takes punishment personally" (2004, p. 984).

Discourses of Criminalization and Piracy

John Gantz and Jack Rochester (2005), authors of *Digital Millennium: How the Intellectual Property Wars Damage Our Personal Freedoms, Our Jobs, and the World Economy*, address one of the more unfortunate outcomes of increased use of peer-to-peer technologies and other file-sharing programs: that **young people and students who use such technologies are increasingly represented as thieves and as "pirates."**

This preemptive criminalization of young people often focuses narrowly on legal, procedural, technical, and academic grounds, ignoring social and rhetorical context, audience, and the social construction of information. For example, Jack Valenti (2003) argues that

> Students operating off their university's broadband, high-speed, state-of-the-art computer networks have a merry old time uploading and bringing down movies, all without paying for them and all with fine fidelity to sight, sound and color [. . .] just a few months ago we learned that one of America's most prestigious and preeminent universities, vexed by the burden of heavy persistent student use of its computer system, actually set up a special server for Gnutella, a well known mightily used site for file-sharing (a discreet description of taking films which don't belong to you) ("A Clear and Present Danger").

Andrew Feenberg's work, for instance, would call attention to the fact that this description defines intellectual-property issues *"strictly in technical terms,"* that it centers around a series of "technical codes" (1999, p. 88). An alternative unpacking of the above description might focus on both its technological *and* its social and discursively defined interests. As Representative Maxine Waters pointed out in a recent congressional hearing on file sharing, for instance, different communities have other interests: "universities aren't going to criminalize America's middle class children. If this were taking place in the inner city, we'd see some movement" ("Educators, Entertainment Industry Team Up").

IN THE TECHNICAL-COMMUNICATION CLASSROOM

For teachers of technical communication, what may be useful in these differing perspectives is the possibility of grounding intellectual-property debates in students' contemporary musical interests and their sophisticated understanding mixing, remixing, sampling, and personal, contextual, moments. Importantly, the perspectives also offer **an opportunity to rethink the ethical dimensions of intellectual property in terms of technical-communication pedagogy and helping students understand that intellectual-property issues may have as much to do with economics, power, competitive-versus-collective forms of information sharing, and corporate control, as with ethics.**

Pragmatically, these competing perspectives characterize the contested environments and some of the complex and politically loaded intellectual-property issues that students will encounter in industry and public spheres after graduation. In such landscapes, they will need informed frameworks for understanding the implications and consequences of their communication practices. Because politics and pragmatism are not mutually exclusive in this world, it makes sense for teachers of the introductory technical-communication classroom to develop activities and assignments that keep these issues at the forefront of students' attention and give them real meaning.

Other similar and related assignments have been suggested by academic scholars who teach communication courses. Among them, TyAnna Herrington's *A Legal Primer for the Digital Age* (2003) provides excellent legal reading and case studies for students and for teachers, and web sites such as Kairosnews (2005) and the Conference on Composition and Communication IP Caucus (CCCC-IP) (2005) regularly provide news, background, as well as activist and research possibilities.

ASSIGNMENT SEQUENCE

The sample assignments included in Appendix B aspire to engage students from a range of majors in their own learning, social, cultural, and professional environments. Many provide opportunities for writing for audiences other than teachers, and they all contain a mix of legal, educational, and professional reading and discussion activities. **A course in technical communication for nonmajors is a particularly compelling opportunity for both students and teachers to work in intellectual-property contexts,** because other than the legislative actions and centralized administrative procedures that I address above, there remain only emerging disciplinary approaches to understanding how to follow the law and how to probe it for developing productive alternatives.

In addition to the activities suggested at the end of this chapter, teachers may want to focus some of their instruction on intellectual property issues around students' majors.

- Students in bioengineering, bioethics, and the social sciences should explore tensions and emerging understandings between intellectual property, traditional, indigenous knowledges, and cultural heritages (WIPO; Alford, 1995; Brush & Stabinsky, 1995; Posey & Dutfield, 1996).
- Students in business and marketing can develop an understanding of work-for-hire contracts (Herrington, 2003), organizational strategies for copyright compliance, and "anti-piracy technologies promoted by media companies" (Burgelman & Meza, 2003).
- Students in education can ground their classroom research and develop an understanding of K–12, community college, and university approaches to qualitative research and copyright issues (McSherry, 2001).
- Students in nursing will benefit from understanding how the scientific research in their field is becoming more expensive to produce, access, and share based on publishers' increasingly centralized control of copyright permissions and agreements (Brown, 2004).
- Students in graphic design, in new media, and in the arts can explore connections between production, ownership, and licensing (Crawford, 1999), and the legal relationship between images and fair use (Aufderheide, 2003; Harper, "Copyright and Image Management," 2005; Thill, 2004).
- Students in music and sound design can create projects based on the historical and contemporary connections between music and copyright (Baran, 2002; McLeod, 2001; Negativland, 2005; *Campbell v. Acuff-Rose Music,* 1994; Sapherstein, 1998).
- Students in literature might be interested to know that John Milton is most likely the first English author to contract a copyright for published work (Lindenbaum, 1994, 1995; Loewenstein, 2002), and they can apply more recent work in copyright and creativity in research or inquiry projects (Lessig, 2004; Negativland, 2005).
- Students in computer science, computer engineering, and the management of information systems can analyze the rhetorical, legislative, and technological implications of peer-to-peer file sharing platforms (DMCA, 1998; Lessig, 2004; Logie, 2003; Oram, 2001; Rheingold, 2003) and reverse engineering (Huang, 2003; Wark, 2004).

Finally, **the best justifications for working with copyright and fair-use materials and contexts in a technical-communication course are in the historical, contemporary, and rhetorical links between communication, law, and**

technology: the first copyright laws emerged as a result of the printing press in late fifteenth-century England; the first major revision to U.S. copyright law, in 1909, was largely a result of contested music-distribution innovations and corporate efforts "to retain their advantages" (Litman, 2001, p. 37), and was amended in 1912 to provide copyright protection for motion pictures (Paterson & Lindberg, 1991, p. 90); the most recent revision of U.S. copyright law, in 1976, was the direct result of photocopiers, computers, and the increasingly decentralized distribution of information and knowledge.

The history of copyright law in the United States is contemporaneous with the history of technology, with communication, rhetoric, and with our efforts to understand the effects and implications of a technological society.

APPENDIX A
Background and Materials for Teachers

This appendix contains assignment sequencing ideas, background, and preparation materials for technical communication teachers. The assignment descriptions here correspond directly to the suggested activities and projects in Appendix B.

BACKGROUND ASSIGNMENT #1:
The Copyright Clearance Center (CCC)
http://www.copyright.com/

The Copyright Clearance Center (CCC) is a good site for analysis and discussion: it began as a publishers' industry group, designed to help collect copyright permission fees from various users. It has since taken on an advocacy and "educational" role, as you'll see within the site. The questions and prompts in Appendix B are intended to guide students through a rhetorical analysis of the site.

Suggested Reading:

§ 107. Limitations on exclusive rights: fair use.
http://www.copyright.gov/title17/92chap1.html#107

Campbell v. Acuff-Rose Music (pp. 92–1292), 510 U.S. 569 (1994).
http://supct.law.cornell.edu/supct/html/92-1292.ZS.html

Selections from Herrington's *A Legal Primer for the Digital Age* (2003), especially pages 93–97 and 106, 107 on copyright and fair use.

Suggested Activities:

It will be helpful to introduce students to Section 107 of the Copyright Law, "Limitations on exclusive rights: fair use." Students should be able to summarize and paraphrase the four factors and give examples of potential noninfringing uses.

Placing intellectual property within these social, cultural, and economic contexts has focused increasing emphasis on some of the emerging and ongoing concerns about intellectual property and its complex economic relationships to technology adoption and use, ethics, and communicators' roles in a global economy.

As a result of these discussions, public advocacy organizations and groups like the Electronic Frontier Foundation have placed intellectual property issues in a different light and have promoted a different kind of globalism:

> Imagine a world where technology can empower us all to share knowledge, ideas, thoughts, humor, music, words and art with friends, strangers and future generations. That world is here and now, made possible with the electronic network—the Internet—with the power to connect us all (Electronic Frontier Foundation, "About").

Corporate and business perspectives offer numerous important sources on the importance of intellectual property. For example, a technical report on peer-to-peer technologies from IBM notes that **"Individuals have new power to reshape the boundaries of control over intellectual property, applications use and knowledge distribution"** (Andrews, 2002, p. 1). This particular perspective is notable for its emphasis on reshaping technological and knowledge boundaries; whereas most business-oriented literature on intellectual property emphasizes control and ownership.

The World Intellectual Property Organization (WIPO) is another important resource in economic and business contexts. In WIPO's case, the literature, materials, and legal contexts are global in reach, and the complexities of intellectual property reveal themselves in high relief in WIPO's language, which is designed to promote IP rights and responsibilities:

> From the classrooms of today will come the entrepreneurs, the scientists, the designers, the artists of tomorrow. WIPO is committed to promoting a culture in which young people can realize this potential. Through well-balanced IP systems and structures, WIPO seeks to help creators across the globe generate economic value from their creations, and so to contribute to the social, cultural and economic advancement of their own societies and of the wider world (WIPO, 2005, "World Intellectual Property Day").

The linking of creativity and innovation with "economic value" underlies almost all of the contemporary debates about intellectual property, and technical-communication teachers have an important role in helping to shape students' awareness and understanding of those debates.

WHAT ACADEMIC RESEARCH TELLS US

If students in introductory technical-communication courses are aware of intellectual-property issues, chances are that they have heard about them as a

Ask students to read and study the CCC site with these questions in mind:

- What kinds of companies and organizations seem to be the intended audience? Corporate? Nonprofit? Educational? How can you tell?
- How and where is the concept of fair use introduced and explained?
- How are the principal rights of copyright holders balanced with the rights of other potential legally protected users?
- Visual analysis: what values are expressed on the site's choice of icons, links, and overall design?

BACKGROUND ASSIGNMENT #2:
The Institute of Electrical and Electronics Engineers (IEEE)
http://www.ieee.org/

The Institute of Electrical and Electronics Engineers (IEEE) "promotes the engineering process of creating, developing, integrating, sharing, and applying knowledge about electro- and information technologies and sciences for the benefit of humanity and the profession" ("About the IEEE":http://www.ieeeusa.org/COMMITTEES/IPC/).

The organization's legislative body, IEEE-USA, maintains a detailed and regularly updated site on intellectual property to promote its lobbying efforts and to update members on legislation and current issues, such as Reverse Engineering. Students should be encouraged to learn about the intellectual-property issues in their field; the IEEE site offers a wealth of materials toward that end and for classroom analyses, reports, and further research.

Suggested Reading (selections from):

Rockman's *Intellectual Property Law for Engineers and Scientists.* Wiley-IEEE Press, 2004. (Especially Chapter One: "Overview of Intellectual Property Law.")

BACKGROUND ASSIGNMENT #3:
Fair Use in Composition and in Technical Communication

This series of short assignments can be used to structure a unit based on copyright and fair-use materials, or to develop increasingly complex group-based projects and presentations.

Suggested Reading:

U.S. Supreme Court Decision, *Campbell v. Acuff-Rose Music* (pp. 92–1292), 510 U.S. 569 (1994), for an example of how the fair-use four factors are adjudicated in legal contexts.

White Paper, "Copyright and Digital Media in a Post-Napster World." Berkman Center for Internet & Society at Harvard Law School. Available online: http://cyber.law.harvard.edu/media/wp2005

BACKGROUND ASSIGNMENT #4:
Peer-to-Peer Technologies, Technical Communication, and Intellectual Property

For this intellectual-property unit, students can be asked to identify and analyze contexts, design, and uses for peer-to-peer file-sharing technologies. Class discussions might include both rhetorical and communication-based approaches to understanding the diffusion of technological innovations (including the pro-innovation bias in most diffusion studies) and students' own experiences with the interfaces.

Students can also read and discuss excerpts from legal documents related to the *A&M Records vs. Napster* case from the 9th Circuit Court of Appeals and other jurisdictions; these materials include technological and rhetorical descriptions of peer-to-peer interfaces, "time shifting" arguments, and the social, cultural, and economic impacts of file-sharing innovations.

Suggested Reading:

A&M Records vs. Napster
http://news.findlaw.com/cnn/docs/napster/napster030601ord.pdf
http://www.law.cornell.edu/copyright/cases/239_F3d_1004.htm
http://www.dml.indiana.edu/pdf/AnalysisOfNapsterDecision.pdf
http://www.eff.org/IP/P2P/Napster/

Oram, Andy (ed.) *Peer-to-Peer: Harnessing the Power of Disruptive Technologies*. O'Reilly, 2001.

Rogers, Everett M. *Diffusion of Innovations*. 4th ed. The Free Press: 1995.

APPENDIX B
Assignments and Resources

ASSIGNMENT #1:
The Copyright Clearance Center (CCC)
http://www.copyright.com/

Analyzing "A Guide to Copyright Compliance for Business Professionals"

According to its web site, the "Copyright Clearance Center, Inc., the largest licenser of photocopy reproduction rights in the world, was formed in 1978 to facilitate compliance with U.S. copyright law." Among the Center's resources is "A Guide to Copyright Compliance for Business Professionals."

> As a business professional, you rely on third-party information to support the work you do. This information is critical to many of the functions—such as sales, market research, communications, strategic planning, new product development, R&D and information services—required to stay competitive and achieve the goals of your organization. Yet, in re-using or distributing content without the permission of the copyright holder, you may be violating copyright law.

> Under copyright law you have a personal responsibility to the copyright holder and to your employer to be compliant. Ignoring this responsibility while you go about your job can put you and your organization at risk. In addition to the legal responsibilities, there are also ethical responsibilities in using content created by other parties.

For analysis and discussion purposes, it is useful to do close readings of the CCC language including, in the passages above, how "business professional," "personal responsibility," and "ethical responsibilities" are combined to establish and inform workplace practices.

ASSIGNMENT #2 :
The Institute of Electrical and Electronics Engineers (IEEE)
http://www.ieee.org/

- Research your university's intellectual property policies or guidelines: Where are they published? What is your university's policy for the copyright of student-produced materials? Who owns the rights to collaboratively produced group projects?
- Analyze the 9th Circuit Court of Appeals task of deciding, in *Sega Enterprises vs. Accolade*, ". . . whether the Copyright Act permits persons who are neither copyright holders nor licensees to disassemble a copyrighted computer program in order to gain an understanding of the unprotected functional elements of the program."
- The court's finding—that reverse engineering is a fair use of copyrighted soft-ware—is supported by the IEEE. Research and report on other stakeholders' reactions to the court's decision. *Sega Enterprises vs. Accolade*, 977 F.2d 1510 (9th Cir. 1992).
- Service-learning opportunity: create an Intellectual Property manual for a local, nonprofit agency that explains fair-use principles and guidelines, how the agency can take advantage of those guidelines and principles, and how to protect against possibly unlawful infringement. A good background source is the Benton Foundation's KickStart Initiative; especially see their Intellectual Property resources, designed to assist teachers, librarians, and community center personnel: http://www.benton.org/publibrary/

ASSIGNMENT #3:
Fair Use in Composition and Technical Communication

Intellectual Property:
Copyright and Fair Use Research Reports

- Contact a newspaper, journal, or book publisher in your field of study and find out its policies and procedures for using their copyrighted materials in various formats: distance-education courses, multimedia production, and educational environments. Can you use images from their site in a hypertext essay or multimedia presentation?
- Form groups of two or three students each, and take opposing sides in an intellectual-property problem. You might have someone represent a digital artist who displays her work on the Web and other parties who claim fair use under varying circumstances; invent the circumstances. Or you might have an online-journal editor argue that students in a writing or communication course cannot use the journal's trademarked logos in their print or online analytic reports.
- Research your school's intellectual-property rules and guidelines, especially as they relate to student-created work. Who owns that work? Do the guidelines account for work created with substantial use of university resources?

ACKNOWLEDGMENTS

My thanks to Cindy Selfe for the seemingly endless patience, encouragement, and multiple responses to multiple drafts—and for asking hard questions.

Pedagogical frameworks like the ones I suggest, assignments that I propose, and collective-learning environments needed to enact them would never be feasible or possible without the active and sustaining support of people who maintain computer- and technology-rich learning environments. For example, I would never have been able to attempt activities related to the analysis and design discussions of peer-to-peer file-sharing platforms without the help of Chris Johnson of the College of Humanities Collaborative Computer Classroom (COHLab) at the University of Arizona; Gavin Brown, Webmaster at Villa Julie College; and Keith West, Dickie Selfe, and their staff of astonishingly supportive consultants in the Center for Computer-Assisted Language Instruction (CCLI) in the Department of Humanities at Michigan Technological University. Because they were willing to try new ideas and approaches to writing-and-media classroom activities, I was too.

REFERENCES

Alford, W. P. (1995). *To steal a book is an elegant offense: Intellectual property law in Chinese civilization.* Palo Alto: Stanford University Press.
Andrews, P. (2002, August 23) Peer-to-peer: More than just downloading music: Executive technology report. Retrieved April 25, 2005, from
http://www-1.ibm.com/services/us/index.wss/mp/imc/xt/a1001245

Aufderheide, P. (2005). The copyright creative stranglehold. *Flow: A critical forum on television and media culture*. Retrieved April 25, 2005, from http://idg.communication.utexas.edu/flow/

Baran, M. (2002). *Copyright and music: A historytold in MP3's*. Retrieved April 25, 2005, from http://www.illegal-art.org/audio/historic.html

Bates, A. (1999). *Managing technological change: Strategies for college and university leaders*. Hoboken, NJ: Jossey-Bass.

Benedict.com. Retrieved April 25, 2005, from http://www.benedict.com/

Bettig, R. (1996). *Copyrighting culture: The political economy of intellectual property*. Boulder, CO: Westview Press.

Burgelman R. A., & Meza, P. (2003). *Finding the balance: Intellectual property in the digital age*. London, Ontario: Ivey Publishing.

Brown, S. J. (2004). Is your intellectual property being pirated? *Journal of Nursing Scholarship, 36*, 290-292.

Brush, S., & Stabinsky, D. (1995). *Valuing local knowledge, indigenous people and intellectual property rights*. Washington, DC: Island Press.

Butler, P. (2004). Much respect: Toward a hip-hop theory of punishment. *Stanford Law Review, 56*, 983-1016.

Campbell v. Acuff-Rose Music (92-1292), 510 U.S. 569 (1994). http://supct.law.cornell.edu/supct/html/92-1292.ZS.html

CCCC Caucus on Intellectual Property and Composition/Communication Studies (CCCC-IP). Retrieved April 25, 2005, from http://www.ccccip.org/

"Circular 21": "Reproductions of Copyrighted Works by Educators and Librarians." United States Copyright Office, 1998. Retrieved April 25, 2005, from http://www.copyright.gov/circs/circ21.pdf

CNI. Coalition for Networked Information. Retrieved April 25, 2005, from http://www.cni.org/Hforums/cni-copyright/

Copyright Act of 1976. Title 17. Retrieved April 25, 2005, from http://www.copyright.gov/title17/

Copyright.com. Retrieved April 25, 2005, from http://www.copyright.com/

§ 107. Limitations on exclusive rights: Fair use. Copyright Act of 1976. Title 17. Retrieved April 25, 2005, from http://www.copyright.gov/title17/

Crawford, T. (1999). Copyright and licensing. *Professional practices in graphic design* (4th ed.). New York: Allworth Press.

Crawford, T. (1999). *Legal guide for the visual artist* (4th ed.). New York: Allworth Press.

Crews, K. (2002, September 30). New copyright law for distance education: The meaning and importance of the TEACH Act. Retrieved April 25, 2005, from http://www.ala.org/ala/washoff/WOissues/copyrightb/distanceed/teachsummary.pdf

Digital Future Coalition (DFC). About us. Retrieved April 25, 2005, from http://www2.ari.net/home/dfc/about/aboutus/aboutus.html

"Distance Ed." Report on Copyright and Digital Distance Education. United States Copyright Office, 1999. Retrieved April 25, 2005, from http://www.copyright.gov/disted/

DMCA. The Digital Millennium Copyright Act of 1998. Copyright Act of 1976. Title 17. Retrieved April 25, 2005, from http://www.copyright.gov/title17/www.copyright.gov/legislation/dmca.pdf

Educators, entertainment industry team up to fight peer-to-peer copyright piracy. *Patent Trademark & Copyright Journal*. 65(2003). Retrieved April 25, 2005, from http://ipcenter.bna.com/pic2/ip.nsf/id/BNAP-5KA85M?OpenDocument

Electronic Frontier Foundation. "About." Retrieved April 25, 2005, from http://www.eff.org/about/

Feenberg, A. (1999). *Questioning technology*. New York: Routledge.

Gantz, J., & Rochester, J. B. (2005). *Pirates of the digital millennium: How the intellectual property wars damage our personal freedoms, our jobs, and the world economy*. New York: Prentice Hall.

Harper, G. Copyright and image management. Retrieved April 25, 2005, from http://www.utsystem.edu/OGC/IntellectualProperty/image.htm

Harper, G. Crash course in copyright. Retrieved April 25, 2005, from http://www.utsystem.edu/OGC/IntellectualProperty/cprtindx.htm

Herrington, T. K. (2003). *A legal primer for the digital age*. Boston, MA: Allyn & Bacon/ Longman.

Huang, A. (2003). *Hacking the Xbox: An introduction to reverse engineering*. San Francisco, CA: No Starch Press.

Kairosnews: A Weblog for Discussing Rhetoric, Technology and Pedagogy. Retrieved April 25, 2005, from http://kairosnews.org/

Kay, K. (1996). Comments on hearings on global competition and Innovation. Federal Trade Commission. January 26, 1996. Retrieved April 25, 2005, from http://www.ftc.gov/opp/global/kay_cic.htm

Lessig, L. (2004). *Free culture: The nature and future of creativity*. New York: Penguin Press.

Levin, B. (2003). *The pirates and the mouse: Disney's war against the counterculture*. Seattle: Fantagraphics Books.

Lindenbaum, P. (1994). Milton's contract. In Woodmansee & Jaszi (Eds.), *The construction of authorship: Textual appropriation in law and literature* (pp. 175-190). North Carolina: Duke University Press.

Lindenbaum, P. (1995). The poet in the marketplace: Milton and Samuel Simmons. In P. G. Stanwood (Ed.), *Of poetry and politics: New essays on Milton and his world* (pp. 249-262). Binghamton, NY: Medieval & Renaissance Texts & Studies.

Litman, J. (2001). *Digital copyright*. Amherst, NY: Prometheus Books.

Logie, J. (2003). A copyright cold war? The polarized rhetoric of the peer-to-peer debates." *First Monday*. Retrieved April 25, 2005, from http://firstmonday.org/issues/issue8_7/logie/index.html

Loewenstein, J. (2002). *The author's due: Printing and the prehistory of copyright*. Chicago, IL: University of Chicago University Press.

Markel, M. (2000). *Ethics in technical communication: A critique and synthesis*. Westport, CT: Ablex Publishing.

McLeod, K. (2001). *Owning culture: Authorship, ownership, and Intellectual property law*. New York: Peter Lang Publishing, Inc.

McSherry, C. (2001). *Who owns academic work: Battling for control of intellectual property*. Cambridge, MA: Harvard University Press.

MPAA. Motion Picture Association of America. Retrieved April 25, 2005, from http://www.mpaa.org.

Negativland. Retrieved April 25, 2005, from http://www.negativland.com/

Oram, A. (Ed.). (2001). *Peer-to-peer: Harnessing the power of disruptive technologies*. Sebastopol, CA: O'Reilly.

Paterson & Lindberg. (1991). *The nature of copyright: A law of users' rights*. Athens: University of Georgia Press.

Peters, M. (2001). Statement to the Senate Committee on the Judiciary. Retrieved April 25, 2005, from http://www.copyright.gov/docs/regstat031301.html

Posey, D., & Dutfield, G. (1996). *Beyond intellectual property: Toward traditional resource rights for indigenous peoples and local communities.* Ottawa, Canada: International Development Research Centre.

Rheingold, H. (2003). *Smart mobs: The next social revolution* (Reprint ed.). New York: Basic Books.

Rockman, H. B. (2004). *Intellectual property law for engineers and scientists.* Hoboken, NJ: Wiley-IEEE Press.

Sapherstein, M. B. (1998). The trademark registrability of the Harley-Davidson roar: A multimedia analysis. *Boston College Intellectual Property and Technology Forum.* Retrieved April 25, 2005, from http://infoeagle.bc.edu/bc_org/avp/law/st_org/iptf/articles/

"TEACH Act." Technology, Education and Copyright Harmonization Act. Retrieved April 25, 2005, from http://www.ala.org/washoff/teach.html

Thill, R. (2004). Intellectual property: A chronological compendium of intersections between contemporary art and utility patents. *Leonardo, 37,* 117-124.

Valenti, J. (2003). A clear and present danger: The potential undoing of America's greatest export trade prize. Ashburn, Virginia. Retrieved April 25, 2005, from http://www.mpaa.org/jack/2002/2002_04_23b.htm

Vaidhyanathan, S. (2003). *Copyrights and copywrongs: The rise of intellectual property and how It threatens creativity.* New York: NYU Press.

Wark, M. (2004). *A hacker manifesto.* Cambridge, MA: Harvard University Press.

Webb, C. (2003). *RIAA vs. the people.* WashingtonPost.com. September 9, 2003. Retrieved April 25, 2005, from http://www.washingtonpost.com/ac2/wp-dyn/A47297-2003Sep9

WIPO. World Intellectual Property Organization. http://www.wipo.int/

World Intellectual Property Day: Message from the Director General. Retrieved April 25, 2005, from http://www.wipo.int/about-ip/en/world_ip/2005/dg_message_05.html

SECTION 2

Focus on Sociocultural Understanding

CHAPTER 6

Preparing Students for Service-Learning Contexts with Case Studies, Scenarios, and Workplace Writer Studies

Gerald J. Savage and Teresa Kynell Hunt

> **OUTCOME:** Students will learn to analyze the contexts of writing tasks within organizational settings.

For reasons that might make an interesting study, audience analysis seems to be the most thoroughly schematized and even technologized of the rhetorical touchstones in technical communication. We suspect that it is the development of formal, detailed schema for analysis that has sustained the dominance of audience in technical-communication practice.

We lack, however, the same detailed schema for the analysis of some other important factors in technical communication—among them, *context*. We lack structured criteria for analyzing communication contexts and the ways they influence communication tasks and products. In this chapter, we will examine how teachers can apply case-study methodologies from industry and service learning from academia to strengthen the analysis of context in technical-communication courses.

But why do we believe a greater emphasis on contextual analysis and contextualizing frameworks for writing is needed?

WHAT BUSINESS AND INDUSTRY TELL US

Professionals who have learned not only why an understanding of context is important but how to analyze it are likely to be particularly valued by

employers. In industry, the ability to analyze and work knowledgeably within the cultural and political factors in an organization can be critical to the success of projects and personal careers. Rachel Spilka cites a 1989 report claiming that job applicants for technical-communication positions who have "training in under-standing the social context or culture of an organization are especially competitive on the job market" (Spilka, 1993, p. 212). The ability to analyze workplace situations is called for in the frequently cited "SCANS Report," the Department of Labor report on changing requirements for worker education in the twenty-first century. "The SCANS list combines complex intellectual analytical skills to be taught within complex situations that integrate the SCANS competencies" (Garay, 1998). Beverly Sauer says, "When industries complain that U.S. schools fail to teach workers the 'literacy skills' needed for safe performance in the workplace, they are frequently describing a larger, contextualized notion of literacy that includes the ability to read and interpret scientific information in a specific context" (Sauer, 1998).

Usability Testing and Context Analysis

The field of usability studies has for many years made use of *cases, or scenarios,* as a way of replicating workplace contexts and tasks. The method is recommended in two of the leading handbooks used in academia and industry (Barnum, 2002; Rubin, 1994). We have observed an industrial usability laboratory where details as minute as the pictures on the walls and materials on the desk replicate those in any office in the company. This company also uses portable usability labs to conduct studies in its offices around the country in order to understand not simply a user's or reader's responses to a software application or procedural document but the workplace contexts in which those materials function. In that company, the application developers and writers of the products being tested must participate in designing the tests, observing the usability tests, and writing the test reports—a sometimes upsetting experience for product creators who hadn't previously considered context and usability.

Business Cases, Use Cases, and Context Analysis

A common genre for analyzing contexts for potential action or change in business is the *business case* **Consultant Nancy Maluso describes a business case as a story: "The one place where all relevant facts are documented and linked together into a cohesive story.** This story tells people about the what, when, where, how and why of the reengineering effort" (Maluso).

In the field of software development, practitioners employ a technically sophis-ticated approach to requirements analysis called the *use case*. Consultant David M. Rubin describes use cases as similar to "classical plays" and to stories, in other words, narratives, which seem to be a universal human strategy for making meaning by connecting it to contexts. Use cases focus on stakeholders, goals, actors, actions, responsibilities, and scenarios. Use case analysis is valued in a number of software development companies and systems departments where we find that it is often relied upon not only by the developers but by interface designers, technical communicators, and customer liaison staff. As Alistair Cockburn explains,

> Use cases help structure project planning information such as release dates, teams, priorities, and development status. . . . The use cases act as the hub of a wheel . . . and the other information acts as spokes leading in different directions. It is for this reason that people seem to consider use cases to be the central element of the requirements or even the central element of the project's development process (Cockburn, 2001, pp. 14-15).

One element of use cases is the usage narrative, "a single, highly specific example of an actor using the system" (Cockburn, 2001, p. 17). Another element is the scenario, which is based on a full use case, providing a kind of playing out of the case in a particular task context in story form (Kulak & Guiney, 2004, p. 48).

What may be particularly compelling to doubting students is Cockburn's (2001) point that use cases must often be read and used by nonspecialist readers. "Simple text is, therefore, usually the best choice" (p. 1). Indeed, "A well-written use case is easy to read. . . . [However,] learning to write a good use case is harder" (p. 2). Equally important for teachers is the fact that the use case, which is a type of context analysis, is just as important as audience analysis in many industry settings.

WHAT ACADEMIC RESEARCH TELLS US

Although context has received considerable attention in studies of professional communication for at least 20 years, it lacks the special status accorded to audience in technical-communication textbooks. Contemporary textbooks provide little guidance for methods of context analysis. This is not surprising considering the complexity of organizational and social contexts. What we commonly find in textbooks are case studies—this genre has long been a staple of textbooks in our field. Case studies seem to function in two basic ways, which sometimes overlap. Perhaps the most common use of case studies is to provide examples in which students are asked to assume the role of the fictional technical writer (Kynell & Stone, 1999; Olsen & Huckin, 1991). The other use is exemplary rather than participative, providing brief narratives about professional writers and writing tasks (Alred, Oliu, & Brusaw, 1992; Kennedy & Montgomery, 2002).

Rachel Spilka, a leading researcher and consultant, has argued that workplace writers are expected to act in two contrasting roles, "responsible for adapting to and perpetuating workplace culture, but also for working toward changing and improving workplace practices. The technical communicator, therefore, has become known as an agent both of social accommodation and social change, or innovation" (Spilka, 1993, p. 207). This view suggests that client-based assignments, or service-learning projects *by themselves* are not sufficient preparation for professional practice. Without learning critical analysis of contexts, students will not be prepared to understand how the text they are trying to write can or should function in its organizational setting. Lester Faigley stated this problem 20 years ago, calling for writing research focusing on "the social perspective." He argued that writers need to be able to "view written texts not as detached objects possessing meaning on their own, but as links in communicative chains, with their meaning emerging from their relationships to previous texts and the present context" (Faigley, 1985).

A number of studies of workplace contexts have helpfully complicated our understanding of the social, organizational, political, and ideological dynamics in technical communication. The challenge for teachers, however, is to translate theories of context into teaching methodologies. We believe we can most effectively achieve this goal in a three-stage learning process. First, we provide students with carefully designed cases for analysis using analytic models we provide; second, we provide opportunities for observation of real communication contexts; and third, we provide them with opportunities for writing in nonacademic situations.

The move to service learning both in composition and technical communication has been an important next step in the pedagogy of context, and yet this approach can too readily plunge students into overwhelming complexities characteristic of "real world writing." We know, of course, that service learning can be effective and can have multiple benefits. For example, Sapp and Crabtree, in their 2002 "A Laboratory in Citizenship: Service Learning in the Technical Communication Classroom," argue persuasively that "service learning provides technical communication students not only with skills and experience for résumé and portfolio building, but this underutilized practice is also a worthwhile component in serving many universities' missions to prepare students to be responsible community members" (p. 412). But for consistent success, there is evidence that service learning requires more faculty time than many teachers may be able or willing to commit (Grabill, 2004). In our own courses we have found that service learning is not the most effective means for introducing contextual learning. Students without adequate preparation in analyzing social and organizational factors are likely to quickly turn their frustration into blame for their peers, their clients, or their instructor.

We have concluded that simply immersing students in workplace writing situations without preparing them with critical tools for analyzing tasks, practices, and social contexts results in students focusing primarily on product development because of the naïve assumption that what the client wants is simply a document (polished and professional in appearance, of course). They do not easily understand that the document must emerge from the organizational context and a particular rhetorical situation; that a number of contextual forces—economic, technical, political, and material—have converged in a particular moment and place resulting in a need for a text. That **text is not an end in itself but the particular means of either sustaining or transforming the context out of which it arises.** Students need to understand not only *what* the text they produce is intended to do but also *why*, in that situation, the text they are asked to produce is the best way to do it.

We believe courses involving service learning could benefit significantly from assessment strategies that place a greater emphasis on contextual analysis and contextualizing frameworks for writing (Carter, Anson, & Miller, 2003). In the sections that follow, we describe how teachers might approach this challenge.

ASSIGNMENT SEQUENCE: FOCUSING ON CONTEXT

Before ever attempting service-learning or client-based assignments, we assign what we call the Writer Profile Report. Students submit a narrative that details what career path they plan to pursue (or, in the case of undecided students, what career path seems most appealing), how they settled upon that choice, what kinds of writing they expect to encounter, and what kind of training/education they believe will best prepare them for workplace writing. These narratives are always fascinating because undergraduate students are typically naïve about the realities of their choices. We use these narratives to prepare a "master list" of career choices in the class, give a copy to everyone and then discuss with them some of the writing responsibilities noted in selected narratives. Students are then required to find a professional in their own field willing to be interviewed and observed, and who will allow the student to see samples of her workplace writing.

The goal of the assignment is for students to learn about the kinds and uses of workplace writing done by someone in the student's intended professional field. We provide guidelines for interviewing and observation designed to elicit a (relatively) detailed description of the rhetorical context of a workplace. This assignment does not in itself provide models based on what Spilka calls "ideal practice" (1993, p. 210). It is certainly informed by rhetorical principles, asking students to find out about the genres, audiences, subject matter, and purposes of the writing their informants do in their jobs. The methodology is ethnographic, which tends to look at the lived, everyday experience—the culture—of the informant.

The **assignment's goal is to enable students to observe the social, cultural, and rhetorical dynamics that the professionals themselves may not be aware of or may be unable to clearly articulate because they are so deeply immersed in their situations that all of the factors seem natural and inevitable to them.**

Analysis and Critique of Professional Documents

It isn't difficult to prepare students to obtain good ethnographic interviews. But they do need practice in describing and analyzing writing situations. Starting a week or so before their workplace observations, we teach some basic rhetorical principles and also outline some concepts suggested by some key workplace communication studies. Because the majority of students in the introductory course have had little if any prior exposure to rhetorical principles, we keep the analytic framework fairly simple. The model of the rhetorical triangle presented by Charles Campbell (1995) in "Ethos: Character and Ethics in Technical Communication" and James Paradis' (1991) rhetoric of the operator's manual enable students to elicit features of texts and their rhetorical situations. They apply these principles in analyzing and critiquing real documents that the students themselves bring from situations they are familiar with—their jobs, their hobbies, their communities, and their social and cultural organizations.

Analysis and Critique of Professional Communication Contexts

Using published case-study research reports, we teach some principles for analysis of workplace communication settings and situations. We have discussed such contextual issues as document cycling and accountability networks, based on the study of writing in an R & D organization by Paradis, Dobrin, and Miller (1985); ownership of workplace texts, based on Dorothy Winsor's (1993) study of engineers and public relations writers; and the potential of "interestedness" for distorting information, based on Hynds and Martin's (1995) study of city engineers presentation of a well-drilling plan to a minority community. In all of these studies, the intertwining roles of organizational structure, organizational goals, authority, and texts are illustrated dramatically for students.

Initially, students apply these concepts to case studies such as those in Kynell and Stone (1999), and to narratives such as those in Savage and Sullivan (2001). We spend several weeks on these activities, along with in-class exercises and first drafts of typical technical communication writing assignments.

Writing in response to case studies or scenarios is an excellent way of placing students in a contextualized task. In responding to scenarios, however, they must consider the characters' personalities and interests, and arrive at decisions about what and how to write in response to each problem. Some current textbooks contain actual cases such as the Challenger disaster and Three Mile Island, but those cases ask the student to occupy the role of a "seasoned" professional who has an understanding of writing and of technology that most students lack. Scenarios are written specifically for students and although they cannot replicate all of the complexity of real situations, the effectiveness of this approach is confirmed in pedagogical research (Skelton, 2002) and in industrial applications. The instructor's manual to *Scenarios for Technical Communication: Critical Thinking and Writing* includes this note:

> Scenarios, while certainly not a perfect solution to issues of audience and appropriate level of technical information, shift the audience analysis away from the instructor and towards the characters presented in each scenario. During research for this text, students regularly talked about different characters mentioned in the scenario as their readers, debated the characters' reactions to different techniques and tactics, and began to think of the technical communication situation as opposed to the assignment, an important distinction in our minds (Kynell & Stone, 1999, p. I).

When they submit final responses to problems, they also submit a "solution defense" (see Appendix B), a written rationale for their choices. When students must justify choices, they think carefully before making decisions. The scenario in Appendix A, taken from Kynell and Stone's (1999) *Scenarios for Technical Communication: Critical Thinking and Writing,* offers one example of how students might be engaged to write in a relevant, meaningful manner.

Case studies/scenarios encourage students to think about social, professional, and personal values that are part of and may problematize any writing situation. Students dialogue collectively about the best ways to solve problems, so they also learn the benefits of collaborative work. They also learn that technical writing occurs across disciplines, and that writing does not exist in isolation from everything else they will face in the workplace.

The Writer Profile Study

How does a context-based approach play out in workplace settings? For the observation of workplace writers and the writing of the Writer Profile Report (Appendix B), we assign students to find a writer on their own. The constraints are: the writer should not be a friend or relative of the student (the rationale being that the situation should be as unique as possible for the student); the writer should not be someone in the student's present or past workplace (same rationale as previous constraint); and all meetings with the writer must occur in the writer's workplace (the rationale being primarily legal and ethical).

The assignment involves three kinds of data collection. First, the student is to interview the writer about her professional background, the work she does, and the organizational structure within which she works. She is also to observe the writer at work, with special attention to any kind of communication tasks, but also looking at the physical setting and the people and activities around the writer (Appendix C). Finally, she is to examine texts the writer develops, and if possible, to try her hand at writing some of those texts herself. Generally two or three visits to the writer's workplace are needed.

While these visits are going on, we ask students to talk in class about what they're seeing and learning, giving them opportunities to compare situations. We also compare the situations they are observing to the case studies and narratives we have studied. These comparisons enable students to begin to draw inferences about the influence of context on writing practices that are often ostensibly similar. Their final report is both an ethnographic narrative and an application of the methods of analysis and critique they had previously practiced only in academic exercises.

As participant/observers, students do not have direct stakes in the situations they observe and analyze. However, they realize that they are likely soon to be engaged in practices and settings similar to the one they observed. For example, Alice, a computer science student, observed and interviewed Marvin, a software developer in the systems division of a manufacturing corporation. Alice wrote in her Writer Profile Report:

> When asked about edit cycles for the completed writing, Marvin admitted there was no such thing in place. He referred again to the time constraints they have placed upon them, which can be a negative aspect of the computer field. . . . Technical writers in Computer Science are restrained and cannot keep up with the fast production of software. This prevents the product from being effectively used by the customer. . . . The answer seems like it would be easy, and that is to

write more documentation in the beginning. . . . It seems that the users would appreciate more explanation and the computer professionals would benefit by having fewer follow-up calls after the distribution of the software.

This analysis, based on Paradis, Dobrin, and Miller's (1985) study, raised professional concerns for Alice because she would be interning the next semester in the company where she did her writer profile study. Marvin had talked about the importance of understanding the audience he was writing for, but demonstrated little practical knowledge of ways to communicate effectively with that audience. Alice concluded that Marvin could actually achieve greater control of his own professional "destiny" by keeping a daily personal log of his projects, which she saw as a way to help him "predict and communicate his objectives for the future": essentially a continual invention process suited to the pace and pressures of Marvin's work environment.

When students move on to a service or client project later in the course, they are able to apply the analytic criteria they have learned to a writing context in which they have real stakes. At this point, they are always looking for guidelines and information to help them organize and assess their project as it progresses. With the principles of context analysis described above, they find that they have the practical guidelines and tools needed to navigate the complexities of a "real world" communication task.

CONCLUSION

The "real world" approach to writing through service learning no doubt helps students to situate their writing practices in the context of the workplace. **But students need preparation for contextual learning. Scenarios and case studies can provide a range of situational challenges that they might never consider otherwise. Indeed, in certain programs, students may not have an opportunity to take advantage of the best "real world" avenues to writing because real clients for service-learning projects may not be available.**

Workplace writing studies and scenario-based models enable effective assessment of student learning because evaluation can be tied back to students decisions about the relationships of contexts to genres, formats, and styles. If programmatic assessment is predicated upon our students' ability to apply knowledge to new situations, then the contextual approach to technical communication will provide a variety of mechanisms for demonstrating those competencies.

As Carter, Anson, and Miller (2003) note, institutional contexts are changing with the revision of first-year writing programs, the introduction of writing- and communication-across-the-curriculum programs (WAC/CAC), and the new focus on outcomes assessment across higher education, not to mention changes resulting from budget cuts and calls for greater educational accountability (p. 102). They also caution that the "major disadvantage of the centralized, heterogeneous course is its isolation from technical contexts" (p. 104). And complicating that disadvantage is the question of whether or not those technical contexts emerge from an academically informed, theoretical base or if they emerge from industry.

Teaching the service course in technical writing, then, means first establishing where and how the course is situated, evaluating carefully the variety of students likely to take the course, and finally determining the kinds of outcomes most associated with the needs of those students. **In order to prepare those students for the workplace, the service course must incorporate context and critical thought.** That, obviously, can be done in a variety of ways. **Of the possible alternatives, the best approach, from our perspective, is to combine case studies and scenario analyses, workplace writer observation, and service-learning models to provide a framework for analyzing context, This combination of approaches provides not only a critical understanding of context, but also opportunities for both self-assessment and programmatic assessment.**

APPENDIX A
Case Study: The Rapidly Inflating Air Bags[1]

You walk into the crash test laboratories of Rhinex Testing Inc., a company that has employed you as an automotive testing engineer for five years now. You spend the bulk of your time evaluating the safety features of contemporary automobiles, including built-in child restraint seats, adjustable safety belts, improved auto safety glass, reinforced bumpers, etc. for one of the top three automotive manufacturers in the country. You like your job. You enjoy crash-test trials, get along with your colleagues, and with two children of your own, relish the opportunity to play a role in safer driving.

You get what will be the first of several cups of coffee, and notice your colleague, Marta Richmond, frowning at the microwave oven.

"What's wrong, Marta?" you ask. "Having trouble nuking that hot chocolate?"

"No," she replies smiling. "I'm not frowning at the microwave oven. I'm frowning at the findings John gave me from yesterday's testing on those new air bags. Have you seen the preliminary report?"

"I just got here," you reply. "I really don't know much about John's work on that, but I did some air bag testing myself about a year ago. What's the problem, anyway? Air bags are mandatory equipment now."

"Yeah, I know, but the crash testing John did on infant and kid dummies was not good," Marta responds, stirring her steaming hot chocolate.

"I don't know how you can drink that stuff this early in the morning, Marta." Shrugging your shoulders, you add: "Now what's the deal with John's testing?"

---

[1] Adapted from Teresa C. Kynell and Wendy Krieg Stone. *Scenarios for Technical Communicators: Critical Thinking and Writing.* Boston: Allyn & Bacon, 1999, pp. 49–52.

"Well," Marta replies taking a deep breath, "three out of the five tests John did yesterday on the infant and small kid dummies pointed to their death from suffocation or broken necks in low speed crashes." Pausing to rinse off a spoon, she adds: "The air bags inflated so fast that the dummies were wrenched backwards with incredible force."

"You know, when I did air bag tests a year ago, we only crash-tested adult dummies," you reply.

"Well, most people keep their kids in the backseat, but a recent case of a child death in Denver caused the manufacturer to contact us for testing," Marta added. "We've tested at speeds ranging from 5 to 35 miles per hour. We've tested infant dummies and kid dummies from two to eight years of age. The six-eight year old dummies were injured regardless of impact speed. The infant and younger kid dummies, though, were killed nearly every time, again regardless of speed."

"Wow," you respond to the numbers. "You know, the home company needs to know about this, and they aren't going to like it."

"Tell me about it," Marta quickly adds. "John and I have to write the home company and report our findings. Fortunately these new, faster-inflating air bags haven't been installed in any cars available to the public. Still, the news isn't good. The big guys were counting on these new bags as a public relations feature of the new line. It wouldn't be so bad, but we're both relatively new here, and we don't begin to know how to report such bad news. This could mean a delay in shipping until the problem is solved."

You pause for a moment and consider the possibilities. Marta is right that the home company won't like the findings she and John put together. It's also true that they are new and could use a little help dealing with the "big guns" who won't be happy.

"I'll tell you what, Marta," you respond. "I've had to deal with bad news before. Remember when the side-mounted gas tanks caused so many problems? I wrote that letter. I'll help you and John write this letter. I think I have enough information to draft a letter that the two of you can then review and revise. What do you think?"

"Oh, God, thanks," Marta exclaims. "We were so worried about that write-up. What can we do to help? What else do you need to know to write the letter?"

You stop for a moment and think. "Well, I need to know the models you tested. Also, is the inflation speed the same on all models?"

"Yep," Marta quickly responds. "We tested four vehicles: Excursion XRZ (a four wheel drive, sports utility vehicle), the Illusion LF (a four-door luxury vehicle), the Roundabout (a two-door, compact style vehicle), and the Nadir 100 (a mid-size, two-door family vehicle)."

"And the results were similar in all four models?" you ask.

"Well, yes," Marta answered. "The vehicle didn't seem to matter as much as the speed of impact and the age of the child. Like I said, all the infant and small kid dummies died or were injured in crashes of even ten miles per hour. The bags inflate so quickly and are so large, they virtually suffocate the child. In some cases, the force broke the neck of dummies."

"Does this mean the bags are worthless?" you question.

"Well, not necessarily," Marta quickly answers. "We can recommend that parents put their children in the backseat. Problem solved. Not all parents will do that, I realize. Also, the bags appear to be fine for adults."

"Yeah, but the bags sound like a problem for the passenger side of the car," you add. "This letter is going to take a little diplomacy."

You take away the information you have and draft a letter to the home office, Marshall Motors, Inc. You'll probably write it to the Senior Vice President for Manufacturing, Stephen Youngs, and copy it to the Safety Manager, Roy Nishi, a good friend of yours. Marshall recently settled a lawsuit with a driver over faulty rear hatch latches. Youngs will not be happy about another manufacturing or engineering glitch. You definitely want to communicate the bad news, but you also want to try and find something positive in the situation. Also, you don't want to point a finger at anybody in particular. This one won't be easy.

APPENDIX B
Solution Defense Content[2]

AUDIENCE
- Who requested the document(s)?
- Who "signs off" on it?
- Who will read it?
- Who may read it for another purpose?
- In my organization, what positions do my readers hold?
- What relationship do I have to my readers?
- What information will my readers understand easily?
- What will I have to explain?
- In how much detail will I have to explain complex information?
- What attitude do the readers have toward the document(s)?
- What objections might they have toward what I'm asking?
- Why would they have these objections?
- How do I address those objections in my document(s)?

DOCUMENT TYPE
- What kind of documents (letter, memo, report, etc.) should I write for my reader?
- Why use that type of document?

DOCUMENT PURPOSE
- What, specifically, do I want my readers to do after they receive these documents?
- When do I want them to take action?
- What plan for taking action have I provided?
- For what other purposes could my document be used?

[2] Adapted from Teresa C. Kynell and Wendy Krieg Stone. *Scenarios for Technical Communicators: Critical Thinking and Writing.* Boston: Allyn & Bacon, 1999, p. 5.

DOCUMENT ORGANIZATION
- How did I organize my information?
- Why did I choose to organize it that way?

SOLUTION EXPLANATION
- What, briefly, did I decide to do to solve the scenario's problems?
- What are my reasons behind this solution?
- What other actions did I consider but reject, and why did I reject them?
- What ethical problems did I encounter and how did I work through them?

OUTCOME FORECAST
- What are the most possible reactions to my documents?
- What would I do to address these possible reactions if given a chance to write another document in response?

APPENDIX C
Writer Profile Study
(20% of course grade, minimum 2000 words)

Your assignment is to write a proposal for conducting the interviews, conduct the interview once I have approved the proposal, obtain samples of the writing done by the person you interview, and write a report of the information you obtain.

THOU SHALT NOTS

1) Thou shall not interview a relative. There is a great deal to be learned from this assignment simply by taking advantage of the opportunity to meet and professionally interview someone you have never met before in the field you intend to work in.

2) For the same reason, thou shalt not interview someone who works in the same organization that you work for.

3) Thou shalt not interview by telephone or e-mail.

4) Thou shalt not interview in a home or nonworkplace setting. Arrange to conduct your interview in person at the workplace of the person you interview. The workplace context is an important part of understanding the writing that people do in their professions. This requirement is also intended to protect all parties involved by avoiding situations that may involve or appear to involve compromises of safety or moral integrity. If a situation suggested by your interviewee makes you uncomfortable, please discuss it with me before agreeing to anything. Neither should you ask an interviewee to be interviewed in circumstances that make her or him uncomfortable.

A. Interview and Observation Proposal

Write a proposal in which you explain who you would interview, why you believe it would be appropriate (in terms of this project) to interview these individuals, and what your interview plan is (schedule, key issues, types of documents you believe you could collect as writing samples, and methodology). The purpose of the proposal is to persuade me that your interviews and observations will fulfill the goals of the assignment. You do not have permission to conduct the interviews until I have given written approval of the proposal. I will read your proposal with the following concerns in mind: (1) Do I understand clearly and completely what is being proposed? (2) Is the proposal appropriate for the assignment? (3) Are there other factors that raise concerns in this proposal?

B. Interviews

In this phase you will arrange to interview a nonacademic professional in your major field to learn how they and their colleagues use writing as part of their work. You will learn some interviewing fundamentals prior to conducting this interview.

C. Writing Samples

Workplace documents are sometimes confidential for legal or proprietary reasons. Nevertheless, interviewees may be willing to provide samples of their writing in which the sensitive information is deleted or crossed out.

D. Writer Profile Report

Your report should provide descriptions and analyses based on the guidelines provided for the study. The writing samples should be discussed within the report, but the samples should be attached and referenced as appendixes. The report should be at least 2000 words, not including writing samples or other appendix materials.

APPENDIX D
Interviewing Guidelines for the
Writer Profile Study

BEFORE AND AFTER THE INTERVIEW

1. Make an appointment
   A. Set time and place
   B. Set duration of interview
   C. Tell the interviewee what to expect (topic, purpose).

2. Do some homework on the person's job, workplace, profession (i.e., don't ask questions you should have known the answers to with proper preparation).

3. Prepare written questions.

4. Treat the interview relatively formally. Thus, it would be appropriate to dress more formally than you do for classes. If you present a professional image you will be taken seriously by the person you interview.

5. Transcribe notes immediately after the interview.

6. Ask interviewee if you can call or e-mail back to verify points or fill in missing information. Follow-ups should be brief and specific, not a whole new interview.

INTERVIEW QUESTIONS AND THINGS TO OBSERVE

1. Observe and take notes on the work setting: the organizational context; the physical features of the work environment.
2. What is the relationship of interviewee to others in the organization; his or her position in the organization; who does the interviewee work with; who supervises the interviewee; who does the interviewee supervise; who does the interviewee rely on for advice about writing; who evaluates the interviewee's documents; who uses the documents the interviewee produces?
3. The approximate amount of writing involved in the interviewee's job. Number of documents written/day or week.
4. What kind of documents does the writer produce? Which of the documents he or she produces does he or she consider most important? Which does he or she write most often? Which are easiest/hardest to write?
5. What is a typical writing project? What stages does it go through?
6. How do circumstances influence his or her writing?
7. How do the circumstances in which the documents are read and used influence the writing?
8. Where does the write's information come from?
9. What communication technologies are used to produce the documentation developed?
10. How does the writer know who the audience is for the documents produced?
11. Does the writer collaborate on writing? With how many other people? How does the writer feel about collaboration?
12. Does anyone edit or review the documents before they are considered final?
13. Is usability testing part of the process of document development for any of the documents?
14. What training, education, or prior experience in writing has the interviewee had?
15. What resources does the writer use on the job to help with writing problems: Internet, intranet, dictionary, style guide, grammar handbook, technical writing text book, etc.?

REFERENCES

Alred, G. J., Walter, E. Oliu, & Brusaw, C. T. (1992). *The professional writer: A guide for advanced technical writing.* New York: St. Martin's Press.

Barnum, C. M. (2002). *Usability testing and research.* New York: Pearson Education, Inc.

Brown, S. C. (1994). Rhetoric, ethical codes, and the revival of *Ethos* in publications management. In O. J. Allen & L. H. Deming (Eds.), *Publications management: Essays for professional communicators* (Baywood's Technical Communications Series; pp. 189-200). Amityville, NY: Baywood.

Campbell, C. P. (1995). Ethos: Character and ethics in technical writing. *IEEE Transactions on Professional Communication, 38*(3), 132-138.

Carter, M., Anson, C. M., & Miller, C. R. (2003). Assessing technical writing in institutional contexts: Using outcomes-based assessment for programmatic thinking. *Technical Communication Quarterly, 12*(1), 101-114.

Cockburn, A. (2001). *Writing effective use cases.* Boston: Addison-Wesley.

Faigley, L. (1985). Nonacademic writing: The social perspective. In L. Odell & D. Goswami (Eds.), *Writing in nonacademic settings* (pp. 231-248). New York: Guilford.

Garay, M. S. (1998). Toward a working English for twenty-first-century schools and colleges. In M. S. Garay & S. A. Bernhardt (Eds.), *Expanding literacies: English teaching and the new workplace* (pp. 21-53). Albany, NY: SUNY Press.

Grabill, J. T. (2004). Technical writing, service learning, and a rearticulation of research, teaching, and service. In T. Bridgeford, K. S. Kitalong, & D. Selfe (Eds.), *Innovative approaches to teaching technical communication* (pp. 81-92). Logan, UT: Utah State University Press.

Hynds, P., & Martin, W. (1995). Atrisco Well #5: A case study of failure in professional communication. *IEEE Transactions on Professional Communication, 38*(3), 139-145.

Kennedy, G. E., & Montgomery, T. T. (2002). *Technical and professional writing: solving problems at work.* Upper Saddle River, NJ: Pearson Education.

Kulak, D., & Guiney, E. (2004). *Use cases: Requirements in context* (2nd ed.). Boston: Addison-Wesley.

Kynell, T. C., & Stone, W. K. (1999). *Scenarios for technical communication: Critical thinking and writing.* Boston: Allyn & Bacon.

Maluso, N. (2005). *The business case—Friend or foe?* 1996-2003. Web page. Quality Leadership Center, Inc. Available: http://www.prosci.com/bus1.htm. accessed January 31, 2005.

Olsen, L. A., & Huckin, T. N. (1991). *Technical writing and professional communication* (2nd ed.). New York: McGraw-Hill, Inc.

Paradis, J. (1991). Text and action: The operator's manual in context and in court. In C. Bazerman & J. Paradis (Eds.), *Textual dynamics of the professions: Historical and contemporary studies of writing in professional communities* (pp. 256-278). Madison, WI: University of Wisconsin Press.

Paradis, J., Dobrin, D., & Miller, R. (1985). Writing at Exxon ITD: Notes on the writing environment of an R&D organization. In L. Odell & D. Goswami (Eds.), *Writing in nonacademic settings* (pp. 281-307). New York: Guilford.

Rubin, J. (1994). *Handbook of usability testing: How to plan, design, and conduct effective tests.* New York: John Wiley & Sons.

Sapp, D. A., & Crabtree, R. D. (2002). A laboratory in service learning in the technical communication classroom. *Technical Communication Quarterly, 11*(4), 411-431.

Sauer, B. (1998). Workers at risk: Or, how you read and write might save your life. In M. S. Garay & S. A. Bernhardt (Eds.), *Expanding literacies: English teaching and the new workplace* (pp. 153-176). Albany, NY: SUNY Press.

Savage, G. J., & Sullivan, D. L. (Eds.). (2001). *Writing a professional life: Stories of technical communicators on and off the job*. Needham Heights, MA: Allyn & Bacon.

Skelton, T. M. (2002). Managing the development of information products: An experiential learning strategy for product developers. *Technical Communication, 49*(1), 61-80.

Spilka, R. (1993). Influencing workplace practice: A challenge for professional writing specialists in academia. In R. Spilka (Ed.), *Writing in the workplace: New research perspectives* (pp. 207-219). Carbondale and Edwardsville, IL: Southern Illinois University Press.

Winsor, D. A. (1993). Owning corporate texts. *Journal of Business and Technical Communication, 7*(2), 179-195.

CHAPTER 7

Adapting Communication to Cultural and Organizational Change

Pete Praetorius

> **OUTCOME:** Students will learn that communicating effectively
> within the context of rapidly changing workplaces requires both
> close attention to interpersonal communication and an
> understanding of organizational culture.

I once worked as a technical-writing consultant for a small power-engineering firm. The engineers and technicians at this company designed and built experimental alternative energy systems for the U.S. Department of Energy or the equivalent ministries of other countries. Although the engineers' workloads were high and the pace was fast, my initial impression was that they were unperturbed and happy. This impression was shattered, however, when one day I was treated to a picture of what life was really like at this seemingly tranquil company. The engineers, it turned out, were not happy. They worked for a company that seemed to be going in many directions at once and for a management that took a *laissez faire* attitude toward leading them. Although it might have been normal for a company of this size and industry to be in a state of flux, the engineers complained that they did not feel in control of the changes that were happening around them, nor did they feel that their managers were in control of the changes that the company was going through. The engineers with whom I spoke that day identified communication between them and management as their primary problem, and because I was the "communication guy," they hoped I might help them solve their problem.

I think these engineers rightly identified ineffectual intracompany communication as the root of their problems and frustrations. Although the founder of the company had a mission and corresponding ethical vision, this vision did not seem to shape the company's communication values and practices. Engineers were left uninformed of day-to-day expectations and long-term goals. As a result, they lived in fear, and a climate of unnecessary stress prevailed. The lesson to be learned from this situation, I believe, is that change *will* happen; but for change to be successful, it is best if it is managed in active ways and through responsible communicative efforts. **To manage change responsibly takes a skillful communicator—one who can communicate well both orally and in writing. Instructors of technical-communication classes are in an ideal position to help students understand the relationship between workplace communication and change as well as discern the nuances of communication values and practices that are unique to changing fields, disciplines, and organizations.**

Change is an important aspect of workplaces, and the documents produced in workplaces play a significant role in both initiating and stabilizing the changes that occur. Because of this transformative potential, good writers will take care to produce texts that meet specific needs and contexts. Indeed, one of the chief respon-sibilities of technical-communication teachers may be to help students understand the nature of such rhetorical contingency: because documents must meet specific needs and contexts, a perfect template for future documents does not exist. Moreover, using templates inappropriately to produce documents may well ignore the needs of dynamic, changing workplaces. Teachers can also help students understand how sound interpersonal-communication skills are integral to both graduates' success and their productive relationships with others in workplace settings. Such relationships depend, in most cases, on taking responsibility for the transformative potential of communication within the context of an organization and its culture.

WHAT BUSINESS AND INDUSTRY TELL US

Industry Wants Workers Who Can Adapt to a Changing Environment

Perhaps it is an oxymoron, but change in the workplace has become routine. The Impact Factory (2005), a management consulting firm, points out some contra-dictions concerning workplace change: "Change is good for you. Too much change is bad for you. We need change to keep us stimulated and creative. We need stability and routine to make us feel safe" (2005). Alan Chapman, an organizational consultant specializing in change management, argues that "Planning, implementing and managing change in a fast-changing environment is increasingly the situation in which most organizations now work." Despite the normalness of change, many workers fear change in their workplace more than just about anything else. One industry source advises managers that, "Your people's fear of change is as great as your own fear of failure" (2001). Peter Giuliano, Chair of Executive Communi-cations Group, clarifies this fear, noting that people "don't resist change, they resist the unknown" (Armour, 1998).

Despite both fear and resistance, however, many industry sources agree that **workers *must* cultivate the ability to adapt to a changing workplace environment.** In a 1997 report by the U.S. Chamber of Commerce, the changing nature of the global workplace was identified as one of the biggest challenges facing industry and education in the United States (Worsham, Blakely, & Maynard, 1997). Likening the challenges inherent in global competition to the challenges presented with the launch of Sputnik, the Chamber's authors appeal to educators to better prepare future workers. Unlike its Sputnik-era predecessor, the present Chamber report goes further than simply calling for a more rigorous curriculum: it suggests that the "challenge is not just to improve education but also to ensure that educators know what else is required in the workplace: a positive attitude, a good work ethic, good communication skills, the ability to work with others, and an understanding and appreciation of diversity." In addition to the disciplines listed in school catalogs, the Chamber also recognizes that workers must learn to adapt to a changing workplace environment, and students should be trained in what are often called "soft-skills," "personal competencies," or "foundational skills." Indeed, "adapting to change" is itself such a skill.

Industry leaders agree that two of the most crucial skills that successful workers possess when it comes to adapting to and facilitating change in a workplace environment are solid interpersonal communication abilities and the ability to understand both the workplace ethics and culture that drive organizations. Both of these skills can help to alleviate the fear of the unknown to which Giuliano refers (Armour, 1998). Chapman (2001) advises managers that a good strategy for managing change is to empower workers by encouraging them to help drive change.

Instructors of technical-writing classes have an important responsibility for preparing students to understand the role that written and interpersonal communications play in facilitating and managing change. **Interpersonal relationships and an organization's culture act as variables that influence the rhetorical nature and the transformative potential of workplace texts.** Certainly, texts within an organization, or among several organizations, may share similarities, but because of the unique context of an organization—the dynamic personal relationships that help form organizations and the cultures that drive organizations—subtle differences among documents are not only easily detectable, but highly desirable (Managers Handbook, 1996).

Industry Wants Workers Who Reflect the Culture of Dynamic Organizations.

Change in workplaces is reflected in the texts that are produced by employees and in the interpersonal communications exchanged among employees. For instance, organizational growth often leads to more thorough internal documentation and detailed written policies (Rotin, 2004). Also, as a field or industry matures, the culture that drives related organizational policies and practices will change. These changes may prompt corresponding changes in discussion, representations of a company or organization, and formats and styles of organizational documents in attempts to address the needs of new audiences and contexts (Gale Group, 2004). Because much of what people do on the job entails writing, recognizing the role that documents play in both facilitating and normalizing change within an organization can help employees adapt to their changing environment and deal with the often stressful process of change (Manager's Handbook, 1996). **If people don't recognize that their documents must change to meet the dynamic needs of their organization and clients, the texts they produce will become increasingly less effective and credible.** In the words of Charles Fishman (1997), employees must "change or die." Those who work in organizations must both change with organizations and help (even if not leading) them navigate the changes they take on.

Part of participating in a changing organization involves sharing the cultural and ethical values that guide it. Employees must, as human-resources personnel often say, be a good "fit" (Caudron, 1994; Mai, 2001). Whether or not a job candidate is a good fit for a company often has less to do with an applicant's qualifications and more to do with whether or not the person shares the values that drive internal company decisions. In short, new employees will be most successful when they mesh well with an organization's culture. Management consultant Jeffery Rotin (2004) notes that an organization's culture is "the heart and soul of an organization. It is the company's core ideology: the shared beliefs, values, behaviours and attitudes of its employees."

However, Rotin (2004) says that too often when companies attempt to manage change, they focus on "the more tangible aspects of a transition—the procedures, finances, policies and structures—without considering cultural issues. Yet managing culture can be the key to effective organizational change." **Those employees who are adept at helping an organization respond to changing times in productive ways are called "change agents" (Fishman, 1997).** Fishman says that change agents leverage "their energy, experience, talent, commitment, and connections to make things happen." Fishman also notes that "most change efforts fail" because of poor communication between managers and their subordinates. An American Management Association survey found that poor communication practices of managers can affect an organization's culture and cause moral problems among employees (Armour, 1998). Fishman notes that not everyone will assume the role of change agent, but all members of an organization must be attuned to the changes taking place and the roles that changes play in shaping the organization.

The changing nature of workplaces is mirrored in the ways that documents change in response to dynamic organizational contexts. As a result, students in technical-communication classes must learn more than how to produce generic workplace communications—instructions, e-mails, memos, letters, formal and informal reports, feasibility studies, and proposals. They must do more than duplicate textbook or classroom examples. As helpful as these examples may be, such texts are unable to serve consistently as templates within the unique structure of dynamic and culturally distinct organization (Kreth, 2000; White, 1998). Because every organization has a unique structure and organizational dynamic, as well as multiple audiences and accepted practices, successful workers must be able to discern and adapt to what organization-specific values and practices govern general behavior, document character, and production.

Industry Wants Workers with Effective Interpersonal Communication Skills

In "Leadership, Communication & Change" management consultant Robert Bacal (2005) tells managers that as "a change leader, communication is your primary and most important tool," and that "change leaders need to be reflective and thoughtful about the ways they communicate." The Impact Factory (2005) adds that **to manage change effectively, employees need "to examine how you got to be who you are, define your communication resources and investigate your personal beliefs and values" and develop "advanced interpersonal skills for greater insight into yourself and others."** Interpersonal skills are critical for both managing and surviving change in the workplace. Managers' need for sound interpersonal communication skills are important for dealing with change in the workplace; but workers, too, are expected to participate and guide change (Chapman, 2001).

Although it's important for job seekers to demonstrate that their skills, values, and cultural understandings mesh with a company's, **keeping and advancing in a job require that new employees get along well with co-workers and participate in moving the organization forward** (Silver, 2001). In her short article in the Society for Technical Communication's *Intercom*, Freya Winsberg (2000) makes just this point. Winsberg says that when helping her boss come up with criteria for measuring job applicants, "The top priority was the personal quality my boss called being a 'grown-up'" (p. 17). Being a "grown-up" has less to do with age—though experience helps—and more to do with mature interpersonal communication skills. Working and communicating well with others is essential.

Job announcements between 1992 and 1998 reflect an increasing desire among employers to hire candidates with solid interpersonal skills (North & Worth, 2004). And, according to a survey by the National Association of Colleges and Employers, "While computer and analytical skills ranked high, interpersonal skills topped the list of new-hire skills" (Grab Bag, 2000). Finally, in a survey of 567 full-time employees, interpersonal communication skills were found to be the number-one factor that influences peer perceptions of fellow employees (Benitez, 2002).

When employees communicate with management and among each other to facilitate workplace change, their interpersonal communication skills are critical to success in having ideas accepted and valued (Fishman, 1997). And, according to the authors of the *Managers Handbook*, writing is regaining its prominence in workplace communication, replacing the telephone, in the form of e-mail and FAX. Rotin (2004) agrees with this assessment and asserts that "Communication initiatives should be frequent and continuous throughout the change process, using various channels, including print, e-mail, video and meetings." As Darcy O'Grady, corporate vice president of human resources at Creo, the world's largest independent supplier of prepress equipment puts it, "I don't have enough ink in my pen to underline the importance of communication" (quoted in Rotin, 2004). O'Grady is not simply speaking about the importance of being able to write well.

That employees possess strong writing ability has long been a concern of business leaders. **In a survey done in the mid-1990s by the National Alliance of Business, "70 percent of those [managers] surveyed said they have a difficult time finding workers with writing skills, although 85 percent of them considered writing skills fairly or very important"** (Managers Handbook, 1996). **According to another survey, listening, reading, and speaking are seen as even more important than writing for managers themselves** (Maddox, 1990). This finding suggests that competent communicators in the workplace need to develop their ability to listen effectively. Workplace communication competence requires the ability to get along with others on an interpersonal level. One way to get along with others is to listen with an open mind and be receptive to opposing views. Fishman (1997) points out that "there is information in opposition," but employees need interpersonal skills to ferret out information from potentially hostile sources.

"Employees react to organizational change in a variety of ways—some positive and some negative" (Stark, 2005). Workers must be able to adapt to interpersonal situations as well as to changes in workplace practices and changes in the rhetorical exigencies of workplace communications. In the best cases, employees' understanding of interpersonal dynamics and the culture of an organization guide the rhetorical strategies they use to compose workplace texts, especially in response to changing contexts. Those employees who understand the dynamic relationship between change, interpersonal communication, and organizational cultures fit the description of what Winsberg (2000) refers to as "being a 'grown-up.'"

WHAT ACADEMIC RESEARCH TELLS US

Loel Kim and Christie Tolley (2004) suggest that a changing workplace terrain is the norm rather than the exception. **The unpredictability of the workplace, Kim and Tolley say, "argues for helping to cultivate a mental agility in students that will enhance individual adaptability"** (p. 382). Like their counterparts in business and industry, many academics have noted that the look and rhetorical nature of workplace documents is driven by dynamic and fluid workplace contexts.

Unfortunately, **technical-communication instruction that relies heavily on template-driven assignments is not successful in passing such values along to those students who graduate and become employees.** In "The File Cabinet Has a Sex Life," for example, Lee Clark Johns (1989) discusses how employees, when faced with writing tasks, often turn to filing cabinets that contain bad examples of writing, hoping for a document that can meet their needs. He notes that this happens for two reasons: first, employees know that these documents have been successfully used in the past; and second, managers less attuned to change often insist that a particular format be followed regardless of its current effectiveness. Johns argues that **when employees, and their management, are more attentive to changes within an industry, they abandon the older, file-cabinet models and develop new, increasingly dynamic and responsive models for organizational communication.** These observations have clear implications for teachers of technical writing. We need to educate students about how and why successful writers are attentive to *specific* rhetorical situations, audiences, and context. **We should also impress upon students the need to apply a dynamic sense of genre to technical-communication tasks rather than a static set of tired templates or dated examples.**

Carolyn R. Miller (1984), defining *genre* in this dynamic sense, notes that **genre is depended on "rhetorical actions based in recurrent situations"** (p. 159). Thus, there is no one *correct* form of, say, a business letter that technical-communication teachers should hold up for students. Letters covering similar content for different audiences, for example, may take on very different forms depending upon the expectations of the readers and the relationships between the writer and the readers. The successful workplace communicator needs to recognize that subtle differences in genre expectations exist between different groups and between members within a particular field, depending on when or where they were trained or went to school (Kuhn, 1977).

Examples of how language and conventions change can be found in a variety of places. A neighbor of mine who is an avalanche course instructor told me that she was recently at a meeting with some rangers who work on Mt. McKinley (Denali) in Denali National Park. These rangers, she says, use different workplace terminology (N. Pfieffer, personal communication, July 8, 2004). Apparently, older rangers use different terminology when referring to the snow conditions that lead up to avalanches than the younger, newly trained rangers. In such cases, external changes—in education, disciplinary understandings, science—can lead to changes in internal methods of communication. An employee who fails to note or adapt to such changes may use outdated terminology, produce documents modeled on older formats, or follow dated genre conventions.

Susan G. Thomas (1994) argues that business-writing instructors can help their students make the right rhetorical and genre decisions by focusing on a "'situational' approach to business writing." Thomas insists that focusing on the different variables inherent in each workplace writing situation is more useful to students than providing a "pre-defined 'cookbook' pattern that grossly oversimplifies business situations." Thomas further notes that "students desperately need to learn how to communicate effectively within the highly political environment of the corporate setting." To provide students with real-world writing situations, Thomas advocates collaborative writing assignments that rely on real client-based or service learning situations. Such situations provide students with experiences that require many of the interpersonal skills and dynamic political savvy demanded by contemporary workplaces.

ASSIGNMENT SEQUENCE: CHANGE, INTERPERSONAL COMMUNICATION, AND ORGANIZATIONAL CULTURE

The following series of assignments is designed to provide students with experiences that will help them adapt to organizational change in productive ways and perform in a manner congruent with an organization's goals. **These assignments present students with classroom situations or experiences that help them acquire the "street smarts" they need to adapt quickly and successfully to dynamic workplace genres and contexts.**

Progressive educator John Dewey (1938) suggests that the most effective curricula—at any level of education—provides an "organic connection between education and personal experience" (p. 25). To promote such an educational experience, Dewey advocates the "principle of continuity of experience" (p. 35), in which "every experience both takes up something from those [experiences] which have gone before and modifies in some way the quality of those [experiences] which come after" (p. 35). In technical-communication classrooms we can help students on two fronts with regard to learning from experience; in fact, Dewey says that a primary responsibility of educators is that "they not only be aware of the general principle of the shaping of actual experience by environing conditions, but that they also recognize in the concrete what surroundings are conducive to having experiences that lead to growth" (p. 40).

Teachers of introductory technical-communication courses can provide such experiences through assignments that situate learning within real contexts. They can also help students imagine such experiences through relevant readings and class discussions. Instructors can also teach students how to analyze their experiences—to make sense of their experiences. **Students who are adept at analyzing their experiences and reflecting on their performance will be better able to deal with organizational change because they will have an awareness of the changes taking place and how they react to these changes.** Technical-writing instructors who understand the continuity of change are most successful when they pass along thesis-important understanding to students.

One common teaching method in technical-writing classes provides students with the experience of working collaboratively with others to produce documents unique to a specific organizational situation. Although students often cite scheduling difficulties, differing levels of personal responsibility, and personality clashes as major challenges when working on such group projects, these assignments can provide important instruction in responding to the dynamic demands of tasks involving interpersonal communication. Especially when group projects are coupled with reflective meta-analysis tasks, they can provide students valuable experience working with colleagues and the opportunity to practice shaping documents to meet the demands of specifically situated rhetorical tasks. **The assignments that follow are designed to increase students' skill in interpersonal communications and help improve their awareness of the relationship between interpersonal communications, workplace change, and the dynamic nature of organizational cultures.**

The main ways that I introduce students to interpersonal communication principles that function well within contexts of changing workplaces include the following:

- Readings and discussions
- Working in groups and then writing a reflective letter on their group experience
- A letters assignment (see Appendix A for assignment sheet)
- An e-mail assignment (see Appendix B for assignment sheet)

Readings and Discussions That Promote an Understanding of Interpersonal Communication within a Context of Change

When introducing students to the nuances of interpersonal communication in dynamic workplace situations, I approach discussions and assignments from a practical standpoint and focus on the changing nature of organizations. I emphasize that all workplace communication must meet audience-specific needs and that these needs will typically change on an interpersonal level and as organizations change. I have found that many students don't realize how much or what types of writing they'll be asked to do in the workplace; many more students don't realize that much of their on-the-job writing will require strong interpersonal communication skills. Whether people are writing memos, letters, or e-mails, or they are working with others on collaborative writing projects, workplace writing generally requires sound interpersonal communication skills. Correspondence via e-mail is also important in helping personnel establish and maintain sound interpersonal relationships. Because interpersonal communication skills play such an important role in dynamic workplaces, I focus on their role in document production.

My goal is to help students understand how strong and responsive interpersonal skills produce more effective written correspondence and make people more effective members of writing teams within dynamic organizations. Whether people are change agents or participating in changing organizations, their communication on an interpersonal level can make the difference between smooth and rough transitions. When I'm focusing on interpersonal relationships (organization and industry change, or organizational ethics), I lead discussions on readings on the subject and have students work in groups. Because of the role that interpersonal communication plays in helping organizations through transitions, students should have a grasp of the following interpersonal skills:

- *I* language and *you* language
- Self-serving bias and fundamental attribution error
- Multiple perspective
- Written and unwritten rules

When appropriate, I introduce the above interpersonal skills by having students read original source material. In many cases, however, academic articles fail to provide the basic information that students need. In these instances, I write my own short descriptions and deliver them on the course web site. Once students have read these short descriptions, we discuss interpersonal skills in class. My goal when discussing these skills is to keep the class discussion focused on the students' own workplace and organizational experiences. Focusing on students' past experiences provides for the continuity of experiences that educator John Dewey (1938) described. **When students have opportunities to make sense of their own past experiences with interpersonal communication, they gain a better understanding of how a new set of such skills might be important to their success in dynamic organizations.**

Correspondence Assignments that Promote an Understanding of Dynamic Interpersonal Communication

As Rotin (2004) points out, written correspondence is reclaiming its prominence in workplace communication. E-mail and faxes are replacing the telephone, and because of this, written correspondence skills are also regaining their value. An understanding of interpersonal communication skills is as important to written correspondence as they are to phone calls or face-to-face communication. The following two assignments give students practice in written correspondence: a letters assignment and an e-mail assignment. (See Appendices A & B for assignment sheets.)

The purpose of the letter-writing assignment is to give students practice in writing workplace letters. Although there are plenty of templates of workplace letters available, I emphasize to students that every situation of workplace writing is unique in its specific rhetorical characteristics because no workplace stands still. As a result, every letter should be seen as an example, a snapshot of a dynamic genre in time, rather than as a template. Thus, I advise students to consider what their audience will expect in a letter by way of content, form, vocabulary, and style. I also tell them to keep in mind that the tone they use in their letters can influence their letters' effectiveness.

The purpose of the e-mail assignment is threefold: First, the e-mail assignment acquaints students with the dynamic conversations taking place in their fields' professional e-mail lists. E-mail lists are a good place to monitor changes within a field; they are often on the cutting edge of new information or industry changes. Second, I have students subscribe to an e-mail list as a way of joining a discourse community. The ability to speak and understand the discourse of a field generally takes a considerable amount of time, sometimes years, and this ability is what separates the novices or outsiders to a field from true insiders. Reading the "chatter" that comes across a professional e-mail list (as well as reading a field's professional publications) is a good way for students to become initiated into a discourse community. Finally, I use this assignment as a way to give students an experience in using e-mail and being part of a professional list. To use e-mail effectively in a professional setting requires experience, which one gains through practice and keen observation.

I ask them to observe their chosen lists for the entire term. I have them note what is being talked about; how long a topic, or thread, persists; and how discussions are structured. Throughout the course we discuss their observations.

Assignments that Promote an Understanding of Institutional Dynamics

In addition to ongoing conversations that emphasize the dynamic nature of organization and workplace documents, the primary assignment I use to convey the dynamism of workplace documents is called "The Changing Face of Business Writing Group Assignment." The purpose of this assignment is twofold: it allows students the opportunity to investigate how a particular type of workplace writing has changed over time, and it gives students the chance to practice their interpersonal skills by working in groups to undertake a short investigation, write a report, and present their findings orally to the rest of the class. In short, this multifaceted assignment enables students to better understand the dynamics that are central to producing documents in changing organizational settings. (See Appendix C for assignment sheet.)

The dynamic genre expectations that direct workplace documents are specific both to a given workplace's interpersonal and organizational culture, as well as to a larger field/discipline's culture—and these cultures can vary dramatically over time. What may be considered an acceptable social practice or constitutive rule during one period, or in one organization's history, may be anathema in another. And so an understanding of how and why change occurs within both organizations and fields/disciplines can be important to employees' success.

Change within both organizations and fields is often reflected in the successful communication practices of experienced employees and practitioners. When discussing workplace dynamics with students, I try to emphasize that changes in an organization's or field's thinking—often due to philosophical or technological advances—directly influence what is said or argued in documents, as well as their production. For example, Mark Runquist (1992) illustrates how rhetorical changes in what counts as evidence among glacial geologists can influence how documents are produced. Early glacial geologists, Runquist notes, speculated and theorized about the nature of such features as glacial erratics and moraines. These early scientists' speculations and theories were based primarily on direct observations—observations gleaned from walking over the landscape. Over time, however, if one wished to be considered what Runquist calls a "good scientist" in the field of Glaciology, he or she needed to document their research through the use of increasingly more sophisticated measurements.

Students have found similar changes in style, format, evidence, and disciplinary practices in other fields and in specific organizations by looking at documents or publications over time. For instance, one group of students noted that coroners' death reports in Butte, Montana made increasing use of medical terminology and became more detailed between the 1880s and the year 2000. As the specific field of forensic medicine became more sophisticated, the death reports the members of this field produced correspondingly became more sophisticated. Like Runquist, this group of students found a good example of how documents within a field grow and change as the field does.

I have used this assignment in 12 classes over a period of 5 years in Montana and Alaska. Neither location is home to many manufacturers or other companies with long histories and large libraries of internal documents that may be examined by student researchers. This lack of private workplace documents has not deterred students in my classes from making worthwhile observations. I encourage students to make use of the many government documents that are available online. In addition, I encourage students to look at documents that come from the fields that they plan to enter. In the few cases where students have had access to internal company documents, they usually gained this access through some sort of connection. The ultimate goal of this assignment is for students to see how documents change over time and to research documents from fields that are relevant to their own lives.

It is important to remind students that they are not writing a history of a particular organization. Although the history of a company is often reflected in, and important to, the documents that members produce, students can sometimes lose sight of the fact that **the primary purpose of the assignment is to note how documents change in response to dynamic conditions within an organization.** When students are looking at their set of documents, I encourage them to note changes in content, arguments, writing style, presentation, tone, and other features. As far as content and arguments are concerned, I tell them to pay particular attention to what seems to be important (what is counted as evidence in an argument), what does or doesn't make it into print, and how the writing reflects changes in the field or discipline over time. It is important to emphasize to students that changes within society or a field will typically lead to changes in what is deemed important enough to merit writing about.

It is also important to remind students to pay attention to social or philosophical reasons for changes in a particular set of documents. It is easy for students to focus on changes in typography and layout, which reflect current styles or technological advances. It is more difficult for students to note the sort of changes that come about because of changes in society's or an organization's values. Although it is important to recognize stylistic changes in documents over time, students can gain the most if they can recognize that changes in values will be reflected in an organization's documents. For instance, when the city of Wasilla, Alaska began implementing zoning ordinances, the new laws reflected changing attitudes within the city's population on what is acceptable behavior. Such a change in attitude was particularly noteworthy in this city, whose citizens have a very strong individualist/independent tradition.

After students have noted what changes have occurred to a set of documents over time, I ask them to discuss and speculate why these changes have come about. That is, I ask the students to provide some analysis and interpretation of changing cultural influences in an organization or field. They can observe, for instance, if there has been a maturing in how the documents have been written. I also ask students to speculate about the advantages and disadvantages of the changes they identify—not all change is necessarily good.

Students who learn how to observe the ways in which documents change over time within specific workplace or organizational settings will be better able to fine tune their own documents to reflect current expectations in a dynamic setting. Those who don't acknowledge the link between current thinking within an organization or field and the documents that are produced are destined to retrieve outdated and potentially ineffective document models from the file cabinet.

CONCLUSION

Workplace communication practices—both written and oral—are to a large part governed by ongoing changes within organizations. Documents produced in workplaces play a significant role in both initiating and stabilizing the changes that occur. The competent workplace communicator recognizes the importance that interpersonal communication and organizational culture play in successfully implementing changes within the workplace. When people produce workplace documents, they do so under the dynamic genre constraints generated by their workplaces' expectations about interpersonal communication, the contemporary values of an organization or field, and both written and unwritten organizational rules. Although plenty of models of workplace documents exist, in the real workplace they can only serve as historical examples rather than truly effective templates. Rather than encouraging students to reproduce templates from the classroom or "file cabinet," **we should help students understand that the workplace is a dynamic and changing place, and that their workplace communication—both oral and written—must reflect the unique circumstances of their situation. When students have an understanding of the relationship between communication and ongoing changes within organizations, they will be better able to discern the subtle differences in reader expectations and produce documents accordingly.**

APPENDIX A
Business Letters Assignment

PURPOSE

The purpose of this assignment is to give you practice in writing workplace letters. Much of the writing that is done in the workplace is correspondence; but not all correspondence is the same. For this reason, I'm asking you in this assignment to try your hand at writing six different types of workplace letters. We have discussed in class some of the key attributes of a good workplace business letter. Keep these points in mind as you work on your letters:

- Get right to your point.
- Anticipate what your audience will expect in a letter by way of content, form, vocabulary, and style. (Note: a letter that may be appropriate for one audience may not be appropriate for another.)

- Anticipate how your audience will react to the content of your letter and present your arguments/points in a manner that will elicit the best possible reaction.
- Keep in mind that the tone you use in your letter can influence your letter's effectiveness—tone can make or break your letter.

WHAT YOU MUST DO

Write each of the following letters:

- Job Offer Letter
- Job Acceptance Letter
- Job Refusal Letter
- Resignation Memo
- Complaint Letter
- Inquiry Letter

Important: Each of these letters should be seen as part of a scenario. Thus, write a cover memo to me explaining the contexts for each of your letters. For each letter tell me who the writer is, who the audience is, and what the situation surrounding the letter is.

HOW YOU WILL BE EVALUATED

You will be evaluated on the following criteria:

- How well do the final versions of your letters cater to your identified audiences? Will your letters meet your audiences' expectations? Do your letters fit the genre for the type of letter and situation for which they were drafted?
- Will your letters be understandable by their target audiences? Are you using discourse-specific language (jargon) appropriately?
- Are your letters accurate?
- Are your letters well written—are they free of mechanics and usage errors?
- Do your letters have a clean professional look (not sloppy)?

APPENDIX B
E-mail Assignment

PURPOSE

The purpose of this assignment is threefold:

- The first purpose is to acquaint you with the conversations taking place in your respective field's professional e-mail lists. Before people implement new ideas in their own workplace they often look to see how others fared when trying out something new. E-mail lists are often where such information is reported and are on the cutting edge for new information or industry changes.

- The second purpose of this assignment is to help you to become a full-fledged member of your chosen field. The ability to speak and understand the discourse of a field generally takes a considerable amount of time, sometimes years, and this ability is what separates the novices or outsiders to a field from true insiders. Reading the "chatter" that comes across a professional e-mail list (as well as reading a field's professional publications) is a good way to become initiated into a discourse community.

- The final purpose of this assignment is to give you some experience using e-mail and being part of a professional list. To effectively use e-mail in a professional setting requires experience, which one gains through practice and keen observation.

WHAT YOU MUST DO

Subscribe to an e-mail list (or newsgroup) made up of professionals in your field. Such lists are often sponsored by professional societies or organizations. You may be required to join a professional organization to subscribe to their e-mail list, but most professional organizations have special discounted rates for students, and if you are serious about entering this field, becoming a member of a professional organization is a good first step toward professional affiliation.

For the entire term I would like you to "lurk" on your chosen list and note what is being talked about. Keep a log and keep track of the different topics that are brought up. Note how long a topic, or thread, is discussed, as well as what threads seem to generate uptake (or a response from other list members) and what topics do not. Throughout the course we will discuss how your observations are going. On the last day of class you must hand in a brief written summary (about 500 words) of your e-mail list observations, and give a brief presentation (about 5 min.) of your e-mail list observations.

HOW YOU WILL BE EVALUATED

You will be graded for this assignment on your brief written summary. Your summary will be evaluated on how thorough you are in describing the activity that transpired on your e-mail list during the semester and on how well your summary is written.

APPENDIX C
The Changing Face of Business: Writing Group Assignment

PURPOSE

The purpose of this assignment is twofold: it will give you an opportunity to investigate how a particular type of workplace writing has changed over time, and it will give you an opportunity to work in groups to undertake a small investigation, write a report, and present your findings orally to the rest of the class. Both of these purposes will help you to better understand the dynamics that are part of producing workplace documents.

As fields, disciplines, or organizational cultures mature, the associated work-place documents tend to take on a more professional appearance and become more focused. Moreover, because nearly all workplace documents are collaboratively written and have a historical context, the number of variables that go into producing a document means that no two documents are the same. Recognizing that documents within a company evolve will allow you to produce documents that best meet your audiences' needs and expectations.

WHAT YOU MUST DO

- Organize yourselves into groups of similar academic and professional interests.
- Locate a set of documents that span a suitable amount of time associated with your chosen subject. The amount of time that is suitable is usually determined by the subject area. If you are studying documents related to the computer industry, you may see a lot of changes over just 30 years. If you are studying an established field, such as government or law, you may need to look at 100 or more years of documents; in either case, you will likely look at the same number of documents.

  A collection of documents taken from a long-standing private organization is particularly good, but such documents are not always the easiest to come by. Government documents, published documents, or other documents in the public domain are the easiest to find. However, check your own personal resources or connections; you may dig up the perfect set of documents. Here are some examples of documents that have been analyzed by students in past classes:
  - coroner's reports from the 1880s to 2000
  - calls for comment of proposed networking protocols
  - city zoning ordinances
  - the use of graphics and layout in owner's manuals for Chevrolet vehicles from 1950 to present
  - fifty years of a city's budget—this group looked at the budget every five years
  - safety warnings in consumer product instructions
  - ads directed toward women for automobiles from the 1920s to present
  - user manuals for computer operating systems from 1980 to present
  - purchase orders for a long-running copper mine
  - changes in how accountants report information
  - annual state-of-the-college letters from a college's president to the state's board of regents
  - internal memos from a utility between the 1930s and the 1980s
- Write a group memo to me that explains the following:
  - what type of document you will be studying,
  - why you are studying these documents,
  - what aspects of these documents you will be studying.

- When looking at these documents, note changes in content, arguments, writing style, presentation, and tone (or any other changes). As far as content and arguments are concerned, pay particular attention to what seems to be important (what is counted as evidence in an argument), what does or doesn't make it into print, and how the writing reflects changes in the field or discipline over time.
- After noting the above changes, discuss why (speculation is okay here) these changes have come about. That is, provide some analysis and interpretation. For instance, is there a maturing in how the documents are written? Are things and ideas that are directly referred to in earlier documents only causally mentioned (or not mentioned at all) in later documents? Has the focus of what is looked at or argued about changed over time? Do later versions of the document incorporate modern stylistic features? Are there advantages (or disadvantages) to these changes? (Not all change is necessarily good.) Keep in mind that you are not writing a history of a particular company. Although the history of a company is often reflected in the documents that members produce, your task is to note how documents change. Changes within a field will typically lead to changes in what is deemed important enough to merit writing about.

  Changes in document characteristics over time will also typically occur due to changes in an organization's structure, mission, or values or due to changes in society's values. It is easy to focus on changes in typography and layout, which reflect current styles or technological advances. It is more difficult, but more rewarding, to note the sorts of changes that come about because of changes in society's or an organization's values. Although it is important to recognize stylistic changes in documents over time, you stand to gain the most if you can recognize how changes in values will be reflected in an organization's documents. For example, when the city of Wasilla, Alaska began implementing zoning ordinances, the new laws reflected changing attitudes within the city's population on what is acceptable behavior. Such a change in attitude was particularly noteworthy in this city, which has a very strong individualist/independent tradition.
- Present your findings to the rest of the class in a 15-minute presentation. Make sure that your presentation is a cohesive and coordinated group presentation, not a patchwork of individual presentations. Note: your oral presentation to the class will come *before* you write up your findings in a report. The reason for this order is so that your classmates and I can provide you with feedback that you can use when compiling your final report. It is typical in business and professional settings for team members to present their findings and receive feedback to colleagues before writing up their final report.
- Write up a group-authored report that discusses your findings. Like your presentation, your report should be a cohesive and coordinated group document, not a patchwork of individual reports cobbled together.
- Write an individual meta-analysis memo discussing how this group project went for you. Discuss your role in the group and what you did to help this project reach completion. Also, discuss the roles of the other members of your group and how well, in your estimation, they did with their assigned and accepted duties. If a member obviously worked hard and did a good job, then this is the place to

acknowledge his or her contribution. If a group member, say, failed to make meetings and/or contribute quality work, then note that failing in this analysis as well. This analysis should take the form of a memo to me and be between 300 and 500 words.

HOW YOU WILL BE EVALUATED

This is a group project. You will be evaluated on your written report, but you must participate in the group presentation. Each member of a group will initially receive the same grade, but this grade may be adjusted up or down after I review each group members' meta-analysis memo.

Your paper will be evaluated on the following criteria:

- How well do you analyze the contributing factors that lead to changes in your chosen documents?
- Did you take the easy way out and only look at surface changes, such as layout and design, or did you also exam how changes in an organization's or society's values prompted changes in your chosen documents?
- Is your report accurate?
- Is your report well written—is it free of mechanics and usage errors?

REFERENCES

Armour, S. (1998, September 30). Failure to communicate costly for companies. *USA Today.* p. 1B. Retrieved March 25, 2005, from Proquest.

Bacal, R. (2005). Leadership, communication & change. Retrieved April 2, 2005, from http://www.work911.com/articles/comchan.htm

Benitez, T. (2002, June). Communication nation: Recent study shows that strong interpersonal skills lead to a better reputation among co-workers. *Incentive, 176,* 6(7). Retrieved January 4, 2005, from First Search.

Chapman, A. (2001). Change management: Organizational and personal change management, process, plans, change management and business development tips. *Businessballs.com.* Retrieved March 24, 2005, from http://www.businessballs.com/changemanagement.htm

Caudron, S. (1994, May 1). Team staffing requires new HR role. *Workforce.com.* Retrieved January 25, 2005, from http://www.workforce.com/archive/feature/22/20/54/index.php

Dewey, J. (1938). *Experience and education.* New York: Collier.

Fishman, C. (1997, April/May). Change: Few can do it. Few can sustain it. Few can survive it. *Fast company:* 8. Retrieved April 2, 2005, from http://www.fastcompany.com/online/08/change.html

Gale Group, Inc. (2004). Business and management practices. *T [plus] D, 58,* 26. Accessed March 24, 2005, from LexisNexis.

Grab Bag. (2000, Summer). *Occupational outlook quarterly.* Retrieved December 17, 2004, from http://www.bls.gov/opub/ooq/2000/Summer/grabbag.pdf

Impact Factory. (2005). Change management: Managing change. Retrieved April 1, 2005, from http://www.impactfactory.com/p/change_management_skills_training/issues_945-2103-87619.html

Johns, L. C. (1989). The file cabinet has a sex life: Insights of a professional writing consultant. In K. J. Harty (Ed.), *Strategies for business and technical writing,* (4th ed.; pp. 145-178). Boston: Allyn & Bacon. (Reprinted from C. B. Matalene (Ed.), *Worlds of writing: Teaching and learning in the discourse communities of work.* New York: Random House.)

Kim, L., & Tolley, C. (2004). Fitting academic programs to workplace marketability: Career paths of five technical communicators. *Technical Communication 51,* 376-386.

Kreth, M. L. (2000). A survey of the Co-op writing experiences of recent engineering graduates. *IEEE Transactions on Professional Communication, 43,* 137-152. Retrieved January 26, 2005, from http://icat.snu.ac.kr/review_paper/7.pdf

Kuhn, T. S. (1977). Second thoughts on paradigms. In *The essential tension: Selected studies in scientific tradition and change* (pp. 294-319). Chicago: University of Chicago Press.

Maddox, M. E. (1990). Communication skills needed by first-line managers. *Manage,42*(2), 11-15. Retrieved October 31, 2004, from ABI/INFORM.

Managers Handbook. (1996). *Looking for a few good writers. Managers Handbook 1*(2). Retrieved January 4, 2005, from Business Source Premier.

Mai, R. (2001, July 1). Corporate communications can make you a winner. *Workforce.com.* Retrieved November 20, 2004, from http://www.workforce.com/archive/feature/ 22/29/20/index.php

Miller, C. R. (1984). Genre as social action. *Quarterly Journal of Speech, 70,* 151-167.

North, A. B., & Worth, W. E. (1998). Trends in advertised SCANS competencies: Technology, interpersonal, and basic communication job skills, 1992-1996. *Journal of Employment Counseling, 35*(4), 195-206. Retrieved January 25, 2005, from ABI/INFORM.

North, A. B., & Worth, W. E. (2004). Trends in selected entry-level technology, interpersonal, and basic communication SCANS skills: 1992-2002. *Journal of Employment Counseling, 41*(2), 60. Retrieved January 25, 2005, from ABI/INFORM.

Rotin, J. (2004, October 5). The art of managing culture through organizational change. *Galt global review.* Retrieved April 2, 2005, from http://www.galtglobalreview.com/business/human_resources.html

Runquist, M. (1992). The rhetoric of geology: *Ethos* in the writing of North American geologists, 1823-1988. *Journal of Technical Writing and Communication, 22,* 387-404.

Silver, S. (2001, November). Can you protect yourself from pink slips? *NJ.com: Career wise.* Retrieved January 25, 2005, from http://www.nj.com/careerwise/index.ssf?/careerwise/ html/ articles/pinkslips.html

Stark, P. B., & Associates. (2005). Employee responses to organizational change. Retrieved January 25, 2005, from http://www.pbsconsulting.com/articles/empresptochange.htm

Thomas, S. G. (1994). Preparing business students for real-world writing. *Education & Training, 36*(6), 11-16. Retrieved February 7, 2005, from ProQuest.

White, J. (1998, November). Insights: Embracing change. *Benefits Canada magazine.* Retrieved January 27, 2005, from http://www.benefitscanada.com/content/legacy/Content/1998/11-98/ben332.html

Winsberg, F. Y. (2000, June). Hiring technical writers: Are we looking for the right skills? *Intercom,47*(6), 16-17.

Worsham, J., Blakely, S., & Maynard, R. (1997). Challenges we face. *U.S. Chamber of Commerce: Nation's business.* Retrieved March 24, 2005, from LexisNexis.

CHAPTER 8

Technological Activism: Understanding and Shaping Environments for Technology-Rich Communication

Richard J. Selfe

> **OUTCOME:** Students will understand the technical, economic, institutional, social, and cultural factors that help shape the technology-rich spaces in which they create, revise, design, and exchange texts. They will be able to use this knowledge to improve the spaces that they and others use for communicating.

In the increasingly technological world of the twenty-first century, few instructors would argue against teaching students how to communicate effectively within the digital landscapes of computer-based networks or against helping them develop some form of digital literacy in technical-communication courses. However, identifying the specific forms of *literacy* that students should acquire in connection with information and communication technologies (ICTs) and articulating the precise balance of skills and understandings that should characterize these literacies has proven much more difficult.

This chapter provides a summary of the ICT literacies that business and academic sources have identified as important and provides a sequenced set of assignments designed to help teach these skills in technical-communication classrooms.

WHAT BUSINESSES AND ORGANIZATIONS TELL US ABOUT ICT LITERACIES

In 1993, the Secretary's Commission on Achieving Necessary Skills, sponsored by the U.S. Department of Labor (2000), issued an influential report that identified those skills that employees would need in the new workplaces of the twenty-first century. **The SCANS report, as this document came to be known, contained advice for both educators and employers (Garay & Bernhardt, 1998) and extended the discussion of ICT literacies beyond the performance of simple computer-literacy skills to include an understanding of the technological systems within which literacy practices were enacted.**

Among the five competencies identified by the SCANS report, for instance, was a cluster of skills and understandings that applied directly to communication in technological environments. The report noted that competent and effective employees needed to be able to select, apply, maintain, and troubleshoot the technologies they used, and to understand the complex "social, organizational, and technological systems" within which these technologies were designed, located, and used. Moreover, as the SCANS report noted, individuals should not only be able to "operate effectively" within such systems, but also to suggest "modifications to existing systems" and develop "new or alternative systems to improve performance" (1998, p. 12).

In 2004, the 21st Century Partnership, a consortium of companies and organizations in the United States, built on the foundation of the SCANS report, adding both scope and depth to the discussion. The consortium's members—among them industry giants such as Apple, Cisco Systems, Dell, ETS, Ford Motor Company, Intel, Microsoft Corporation, National 4-H Council, Texas Instruments, Time Warner, and Verizon—have identified four important categories of understanding and performance required of employees in our rapidly changing technological culture:

- **Social responsibility:** "Acting responsibly with the interests of the larger community in mind; demonstrating ethical behavior in personal, workplace, and community contexts."
- **Critical and systems thinking:** "Exercising sound reasoning in understanding and making complex choices, understanding the interconnections among systems."
- **Information and media literacy:** "Understanding the role of media in society."
- **Communication skills:** "*Understanding*, managing and creating effective oral, written and multimedia communication in a variety of forms and contexts" (*my emphasis*).

Within these categories, the *Partnership* report identified the following literacy skills as necessary tools for success in computer-based environments:

- **Analyzing, accessing, managing, integrating, evaluating, and creating information** in a variety of forms and media. Understanding the role of media in society.
- **Understanding, managing, and creating effective oral, written, and multimedia communication** in a variety of forms and contexts.

- **Exercising sound reasoning in understanding and making complex choices,** understanding the interconnections among systems.
- **Developing, implementing, and communicating new ideas** to others, staying open and responsive to new and diverse perspectives.

A concise summary of the advice these two germinal reports have to offer for teachers of technical communication, I would argue, would read as follows:

Students should understand the technical, economic, institutional, and social issues associated with the technological systems they use for creating, revising, and exchanging texts; and, in light of that understanding, they should be able to act responsibly to improve these systems.

WHAT ACADEMIC RESEARCH TELLS US
ABOUT ICT LITERACY SPACES

In academic circles, too, scholars have come to understand technology issues more robustly by examining them within the context of related social, cultural, and economic formations. Intellectuals such as Martin Heidegger, in *The Question Concerning Technology* (1977), and Jacque Ellul, in *The Technological Society* (1964) have offered increasingly critical perspectives on technology, on technology's relationship to the project of science, and on the many changes that these two linked formations have exerted on the lived experiences of humans during the twentieth century. Such work has led, in more recent years, to understanding the crucial role of human agency in technological contexts. Andrew Feenberg in *Questioning Technology* (1999), for instance, has argued that individuals must take on the responsibility of constantly redesigning those technological systems on which they have become dependent. His efforts have suggested that we need more than a critical *assessment* of technology—we need a critical *engagement* with technological systems.

This same sense of contextual understanding, critical engagement, and responsibility for technology, I argue for in this chapter and elsewhere (Selfe, 2004), should extend to our own and students' digital literacy practices in technical communication classrooms, especially those that make use of the computer-based labs, classrooms, centers, and networked online environments. Technology-rich communication environments surround us these days and offer teachers and students limitless opportunities to engage critically in the redesign process that is often going on behind the scenes. Given this situation, we cannot hope to fully understand the practices of contemporary technical communication, and ICT literacies, unless we understand, as well, the technology-rich environments within which they take place, As Donna Haraway sums it up, "the best way to find a 'larger vision' of technology is to be somewhere in particular" (1995, p. 187).

In this chapter, I interpret Haraway's comment to mean that **teaching students to be literate in the digital environments that support technical-communication classrooms involves more than teaching them to read, compose, and exchange communication in ICT environments. It also necessitates that we teach them**

how to investigate and understand the many dimensions of the specific environments within which they perform these actions.

Andrew Feenberg (1999) identifies such systems and environments as "underdetermined" (p. 4) because they are always in the process of being designed or redesigned and because they provide individuals and groups with entry points into that design process. As he notes,

> . . . technological designs are negotiated achievements involving many partners, not [simply] rational inspirations that spring full blown from the mind of an individual genius or pure laboratory research. The design process is the place where the various social actors interested in developing technology first gain a hearing (Feenberg, 1995, p. 4).

We can help prepare students for twenty-first-century workplaces by teaching them how to use their ICT literacy skills to intervene in the design process in order to serve both their own needs and the needs of other computer users.

It is useful to note here that contemporary technical-communication textbooks provide students only very limited advice in this regard. The advice that is given, typically, is narrow in scope—focused primarily on how to design, compose, edit, and exchange texts in digital landscapes. Textbooks seldom address the importance of articulating and analyzing the problems associated with technical systems or of teaching students how to undertake informed efforts to change such systems.

ASSIGNMENT SEQUENCE

The sequence of assignments I describe below (as represented more fully in Appendices A, B, and C) is designed to give technical-communication students an opportunity to investigate several dimensions of the ICT spaces they use for communicating, with the aim of expanding their understanding about what goes on in such sites and of making suggestions for change that would serve a community of communicators. The sequence is flexible enough to allow students to explore a range of computer-supported spaces on campus, in corporations, or in community organizations.

Assignment 1: Researching Technology-Rich Communication Environments

The first part of the assignment sequence involves three tasks: first, students are asked to identify the many technology-rich environments that surround them in their academic, professional, or personal lives. Students (either individually or in teams) then decide on one environment worth studying in some depth. They observe the environments, collect policy and administrative documents about the purpose and workings of the environment, and interview or conduct focus groups about the many dimensions of the environment: After reviewing all their research, students contact important stakeholders with follow-up questions and collect all their materials for the second assignment in the sequence.

Assignment 2: Reporting on Technology-Rich Communication Environments

The second part of the assignment sequence asks students to synthesize what they have learned and author a report aimed at the individuals in charge of the environment they have selected—one which begins with an executive summary and ends with recommendations for appropriate design changes.

Assignment 3: Presenting on Technology-Rich Environments

The third part of the assignment sequence asks students to design a presentation (Web, PowerPoint, etc.) to give to the class *and* to stakeholders from the specific environment they have selected. Students are also asked to create a final portfolio for the stakeholders of the environment they have analyzed and hope to change.

The sequence generally takes 3–4 weeks and requires that instructors and students create and evaluate a number of team project documents: proposals, interview and focus group notes and summaries, concept maps, reports, and presentations.

CONCLUSION

In an increasingly technological world, faculty who teach technical-communication courses have come to recognize that literacy and computer technology are now inextricably linked within many workplace contexts. These same faculty can help students become increasingly productive citizens in such environments by teaching them how to study and shape technology-rich communication spaces. As the following quotation indicates, computers and the communication environments in which they are used will continue to shape the lives of students long after they graduate.

> Graduating students will find that high-paying jobs will require increasing amounts of, and ranges of computer literacy as the ubiquity of computers—and the networks that connect them—affects all aspects of manufacturing and service industries within this country and others. And, in this increasingly wired social context, electronic networks and communication devices will influence not only the people with whom most citizens work, but also the environments within which they interact. Such environments will become the basis for national and international connections . . . (Selfe & Selfe, 2002).

As the SCANS report and the 21st Century Partnership (2004) suggest, it is no longer adequate for students simply to work efficiently and effectively within these spaces. They must also understand the larger social and cultural contexts within which these spaces exist, the people who use them, and their relationships to institutions, organizations, and communities. **We do our best job of teaching technical communication, in other words, when we help students become technology activists (2004, p. 150), productive designers of technology-rich communication spaces and systems.**

APPENDIX A

ASSIGNMENT 1:
Researching Technology-Rich Communication Environments

1. As a group, brainstorm online or in class a list of the many technology-rich communication environments that are important in your life as a student. Think broadly about possible sites, but name them as specifically as possible. Don't forget to consider community, corporate, and organizational spaces as well as campus environments. Consider, among other options, smart classrooms, conferencing facilities, labs, social gathering places (wired coffee shops, for instance), dorm facilities, community centers of all types, local libraries, or K–12 school environments. Make sure that you choose a technology-rich communication space you are familiar with, one to which you have access, and one within which you can identify a willing and knowledgeable contact person.

2. Choose a site for your team to study. I recommend research teams that have a blend of students who have more and less technical expertise. Class members who want to join a research team need to write a memo to the teacher explaining which team they want to join and why, why they want to study and map the infrastructure of the particular environment the team has identified, and why this online or physical space is of personal interest and importance. Each team member should be able to define the community this technology-rich space serves and the community that uses (or could use) the site. After teams have been approved by the teacher, all teams and individual authors will submit a short proposal to the teacher that describes the technology-rich environment they have selected, why it is worth studying, what types of users are served by the space, and who will serve as contact person and advisor for the site.

3. Observe the technology-rich communication environment that you have chosen and take extensive notes on the dimensions listed below. In addition, set up interviews or focus groups with a range of stakeholders in the environment. For instance you might want to interview a "typical" user of the environment, an "advanced" or experienced user, a teacher (if appropriate), the technical administrator of the site, a nontechnical administrator of the site, or a sponsor of the environment, any person who has some stake in the communication environment is an appropriate person to interview. Either set up one-on-one interviews (with one or two interviewers present each time) or set up focus groups with people who can offer diverse perspectives on the various dimensions of the technology-rich communication environment you have chosen. Be sensitive to these people's busy schedules and to the limited time they will have to respond to your questions. Be professional, courteous, and efficient in the time you spend with them.

Be sure to get signed permission forms from everyone you interview. On the permission form you design, be sure to include a brief description of the project and its goals, the purpose of the interview, the ways in which it will be conducted (e.g., by e-mail, by telephone, in person), the ways responses will be recorded (e.g., handwritten notes, audio recordings, video recordings), the ways in which responses will be used (e.g., anonymous or named), and the people to whom the final report will be distributed. In addition, include spaces for a signature and date.

4. Use the following questions to explore the four major dimensions of your chosen technology-rich communication environment. Add or alter questions to follow up on interesting interview responses.

Human-to-Human Relationships

- Who are the primary users of the communication environment? Secondary users?
- What are the typical communication projects that are officially supported within the environment? Are there "unofficial" projects that occur there as well?
- Who are the important support people? How are they recruited, trained, and paid for the support they provide? How many of them are there compared to the number of users of the system?
- Who do users contact when they have difficulties and how is that contact made?
- Who do users contact if they want to make productive suggestions?
- Is there a technical or administrative team that makes decisions about the communication environment and how will it change over time? Who are they and when and where do they make routine decisions?
- Who controls the budget for the environment and is it publicly available?

Human-to-Institution Relationships

- What relation does this technology-rich communication space have to the mission of the organization/business/institution?
- What kinds of institutional/organizational communication happen within this site? What kinds do not happen?
- Can a rough or precise organizational chart be drawn to situate this environment within a larger framework or institutional setting?
- When are fiscal decisions (purchases and hiring) concerning the environment made, and who helps make them?
- Is there an amortized budget for the environment (a budget that indicates replacement costs over time?
- Who or what group makes policy decisions about what can happen in the environment: security, access levels, software and hardware use, procedures, etc.? How, when, and where do these policies get established and changed?

Human-to-Physical Space Relationships

- Describe the workstations available. Why were they chosen? How many people have access to them? How well or how poorly do they support communication activities?

- Who decides on the spatial arrangements of the environment including ambiance, placement of peripherals, types of seating, lighting, etc.? How do those decisions support (or fail to support) communication activities?
- What times of the week and day is this space available?
- Is there support staff available during open hours? If so, what are their primary responsibilities? If they don't know how to fix something, what additional options do users and support staff have?
- Are stakeholders willing to consider changes to the current arrangements that might improve communication or make such communication easier? Who would be involved and when might they make such decisions? Under what constraints will they be operating?
- Who has access to this physical space? Are there intentional restrictions? Are there users who need access but don't have it?
- Is there a designed relationship between the online communication environment and the physical setup?

Human-to-Virtual Space Relationships (if appropriate)

- What primary virtual activities are meant to be supported by the online communication environment? What other types of activities occur in this virtual space?
- Who decided on the arrangement of the virtual environment including the software and storage spaces available?
- What communication output devices or systems do users have access to (printers, DVD/CD burners, online databases, Web space, online management systems) for courses or for projects?
- Who has access to this virtual space? From where do they have access? Are there intentional restrictions?
- Are stakeholders willing to consider changes to the current online system? Who would be involved and when might they make such decisions?

5. Collect detailed notes, e-mail responses, video tapes or audio tapes from your interviews or focus groups. Make sure you have signed permission forms from all of the people you have interviewed. Spend 20 minutes as soon as possible following the event to write down your impressions. In most interviews, you will need to ask follow-up questions.

Interviewing can seem a little intimidating. Consider the following suggestions; they may make the process a little easier, and they will certainly help make the experience more successful.

- Normally the people you interview will be delighted to hear from you and will usually do everything in their power to make you feel comfortable and give you the information you request. To maintain this relationship, be prepared, be professional, be pleasant, and be efficient in your dealings. Make appointments and keep them; come prepared and don't waste time. Tell people everything they need to know about your project. Thank people for their time.

- Your main task is not to get through a list of questions; rather, it is to gather information from informed sources. In this interview situation, your list of questions is simply a backup plan in case the conversation slows down. What you really want is an informal and informative exchange. In a good interview, the person(s) you interview will do most of the talking. Your job is to listen and prompt.
- E-mail may seem like the easiest way to complete this interview, but it often is not. First, it may be too much work for the person you interview to write out extensive responses to all your questions—and to follow-up questions. Remember, you don't want the interview to be a burden for the person(s) involved. You should be the one doing most of the work. Consider other options: phone conversations or instant messaging sessions.
- Promise to send all the people you interview your short informal report and be sure to send each person a thank-you note.

Evaluation Criteria:
Researching Technology-Rich Environments

1. Initial 2-page proposal is well written and includes the environment's name, its value, user description, stakeholder names, and contact information.

```
1                              3                              5
|_____|_____|
```
Not very detailed   Rich in detail
Poorly written   Well written

Comment:

2. Observation notes are detailed, insightful, and cover the multiple dimensions of the environment.

```
1                              3                              5
|_____|_____|
```
Observations one-dimensional   Well developed and complete

Comment:

3. Interview/focus-group notes are detailed and cover the multiple dimensions of the environment.

```
1                              3                              5
|_____|_____|
```
Minimal notes   Well developed and complete

Comment:

4. The final product is appropriate for the audience.

```
1                              3                              5
|_____|_____|
```
Needs work   Excellent

Comment:

5. The final product is professional and mechanically correct.

```
1                              3                              5
|_____|_____|
```
Needs work   Excellent

Comment:

Evaluation: Needs considerable work   Average   Outstanding

APPENDIX B

ASSIGNMENT 2:
Reporting on Technology-Rich Communication Environments

Members of research teams should compare and pool notes and review all the answers and suggestions that came out of the interviews or focus-group meetings. Each team of researchers will write a very short executive summary of findings and then lay them out in more detail, arranged within the four dimensions listed previously.

All reports will contain design diagrams and conceptual maps of the multiple dimensions of the technology-rich communication environment that have been studied. Let me explain: one of the best ways to make sense out of all that your team has learned about the selected site is for each researcher in your group to draw a conceptual map of the facility/environment—one that represents the key elements operative in each dimension as well as the operative forces that constrain change in a facility/environment. Use the best of these conceptual maps (these visual aids) to help explain what you observed and what conclusions you reached in connection with the communication environment you are studying.

Your conclusion section should outline possible changes or improvements that might be implemented so that the community of users can be more effective in their communication within the technology-rich environment. List all the changes that seem beneficial even though some will be more difficult to implement than others.

Edit and revise your report for clarity and for surface errors and hand it in to your teachers for a collaborative grade. Hand in all interview notes, tapes, or records as an appendix to the report.

Evaluation Criteria:
Reporting on Technology-Rich Communication Environments

1. The executive summary parallels the report structure and covers all four dimensions of technology-rich environments.

```
1                           3                              5
|_____|_____|
```
Not very detailed   Rich in detail
Poorly written   Well written

Comment:

2. The conceptual mapping is detailed and connects all four dimensions of the technology-rich environment.

```
1                           3                              5
|_____|_____|
```
Minimal mapping   Well developed and complete

Comment:

3. Report conclusion tactfully outlines productive change and suggestions for improving the technology-rich communication environment for the community of users.

```
1                           3                              5
|_____|_____|
```
Minimal notes   Well developed and complete

Comment:

4. The final product is professional and mechanically correct.

```
1                           3                              5
|_____|_____|
```
Needs work   Excellent

Comment:

Evaluation: Needs considerable work   Average   Outstanding

APPENDIX C

ASSIGNMENT 3:
Presenting on Technology-Rich Communication Environments

1. As a group, decide which changes or improvements you would most like to see in the environment that you have studied. Justify your choice and document it, if possible, with data from your research (interviews, focus-group sessions, and observations).

2. Design two presentations (Web, PowerPoint, etc.): one to give to the class *and* one to give to stakeholders from the technology-rich communication environment. Both presentations should explain your suggestions and how you see those suggestions fitting into the constraints and mission of the environment. Make a presentation to the class and to at least one representative of the environment you have chosen. The two presentations should vary to the extent that the audiences for the presentations vary.

3. Bundle your research data, report, and presentations on change into a final project portfolio. Create appropriate cover art for your project. Make final copies available to both your teacher and to an appropriate representative of the environment that you studied.

Evaluation Criteria:
Presenting a Technology-Rich Communication Environment

1. Changes or improvements are clear and well supported with research data.

1   3   5

Not very detailed   Rich in detail
Poorly written   Well written

Comment:

2. The presentation makes use of data and conceptual mapping. Visual components are appropriate for class, teacher, and client audiences.

1   3   5

Minimal mapping   Well developed and complete

Comment:

3. The final project was made available to the TR representative and is professional and mechanically correct.

1   3   5

Minimal notes   Well developed and complete

Comment:

Evaluation: Needs considerable work   Average   Outstanding

REFERENCES

Ellul, J. (1964). *The technological society* (J. Wilkinson, Trans.). New York: Vintage.

Feenberg, A. (1995). *Alternative modernity: The technical turn in philosophy and social theory.* Berkeley: University of California Press.

Feenberg, A. (1999). *Questioning technology.* New York: Routledge.

Garay, M. S., & Bernhardt, S. A. (Eds.). (1998). *Expanding literacies: English teaching and the new workplace.* Albany, NY: SUNY Press.

Haraway, D. (1995). Situated knowledges: The science question in feminism and the privilege of partial perspective. In A. Feenberg & A. Hannay (Eds.), *Technology and the politics of knowledge* (pp. 175-194). Bloomington, IN: Indiana University Press.

Heidegger, M. (1977). *The question concerning technology and other essays.* (W. Lovitt, Trans.), New York: Harper & Row Publishers.

Partnership for 21st Century Skills. *Learning for the 21st century: A report and mile guide for 21st century skills.* (http://www.21stcenturyskills.org/reports/learning.asp on 3/28/05), 11, 2004)

Selfe, R. (2004). *Sustainable computer environments: Cultures of support for teachers of English and language arts.* Cambridge, CA: Hampton Press.

Selfe, R., & Selfe, C. (2002). Critical technological literacy and English studies: Teaching, learning, action, and response. In R. P. Yagelski & S. A. Leonard (Eds.), *The relevance of English* (pp. 344-381). Urbana, IL: NCTE.

U.S. Department of Labor, The Secretary's Commission on Achieving Necessary Skills: What work requires of schools. Letter to parents, employers, and educators. (http://wdr.doleta.gov/SCANS/whatwork/whatwork.html on 3/28/05) 2000.

CHAPTER 9

Communities of Practice: The Shop Floor of Human Capital

Tracy Bridgeford

> **OUTCOME:** Students will be able to communicate effectively within and among communities of practice, demonstrating understanding of how these communities work, how they establish expectations for membership, and how they agree on rules for negotiating meaning.

Communities of practice are the shop floor of human capital, the place where the stuff gets made.

—Thomas A. Stewart, *Fortune* magazine

Collaboration is a staple of technical communication pedagogy. Most of the major textbooks in technical communication emphasize the importance of collaboration by including chapters devoted to the topic (see, e.g., Burnett, 2005; Lannon, 2003). These chapters typically address aspects of collaboration that we teach: challenges of team projects, methods of collaboration, project and time management, communication with group members, development of active listening skills, and management conflict. In introductory technical-communication courses, teaching the collaborative nature of writing typically covers a majority of the course workload. For instance, typically, students create, design, and produce various technical documents as a group.

To facilitate collaborative environments, instructors establish communities in and out of the classroom within which students learn to participate in concert with others. Although case studies, client-based and service-learning projects, or fiction[1] have resulted in successful collaborations and classroom activity, both industry and academia have begun embracing a more recent concept—communities of practice—first developed by long-time collaborators Jean Lave and Etienne Wenger (1991) and later fully characterized by Wenger (1998) in *Communities of Practice: Learning, Meaning, Identity*. This concept, says Wenger, grows out of a theory of learning that emphasizes participation in practice as a process of learning and knowing that includes a way of talking about practices, shared histories, social network, and personal histories, all of which contribute to the development of an identity of participation. This identity is essential if participants are to progress from newcomer status to experienced member within a community. **Because we do not have direct access to the actual communities students will encounter in their specialized fields, technical communication instructors need to provide students with knowledge about how to enter communities through practice.**

WHAT BUSINESS AND INDUSTRY TELL US

Interest in supporting communities of practice in industry has increased steadily since the early 1990s when researchers Jean Lave and Etienne Wenger (1991) at Xerox's Palo Alto Institute for Research on Learning put forth **a new theory of learning, which they describe as** *legitimate peripheral participation* **(LPP).** Steadily cited since then, LPP refers to **the enculturation process in which newcomers engage when joining a community, a process that makes learning about the community's enterprise meaningful to them.** In communities of practice, LLP provides an entry point into learning, an "integral constituent" in practice (p. 35), and is, for Lave and Wenger, always "an analytical viewpoint on learning, a way of understanding learning" (p. 40). Because meaning is created in the act of participation, members adopt what Wenger (1998) in a later work calls an *identity of participation*, which involves developing an understanding of a community's ways of working (practices), recognizing the expectations for belonging (identity), and learning and agreeing to the rules for negotiating meaning (social network). Although these "communities of practice" (Wenger, 1998, p. 6) have no core membership *per se*, they typically have a "core of participants whose passion for the topic energizes the community and who provide intellectual and social leadership" (Wenger & Synder, 2000, p. 141). Generally, participants are situated at various levels of participation, including full, peripheral, and boundary, all of which contribute to the "diversity of relations involved in varying forms of community membership" (p. 37). This diversity of membership enables new understandings and practices in the mutual engagement of practice.

---

[1] Several scholars, including myself (2004), have used fiction in the technical communication classroom to create contexts (see, e.g., Karis, 1989; Kilgore, 1981; Nelson, 1986), and some scholars have argued against it (see, e.g., Allen, 1989). Unfortunately, most of the discussions about using literature within the technical communication classroom took place at a time when PhD candidates were mainly awarded in literature, creating a defensive response that hindered the continuation of this discussion.

Communities of practice (COP) are "groups of people who share a concern or a passion for something they do and who interact regularly to learn how to do it better" (Wenger, 2004).[2] Together, participants in COPs share work practices and tools, a common language, a knowledge framework, and a way of knowing and understanding practice. Key to understanding how to participate in such communities is how members manage themselves and their enterprise. Participants in COPs actively engage in knowledge building and shared learning experiences that help them build a repertoire of practices. As participants in a COP, members must be fully engaged in the practice, willing to "give as well as take knowledge in order to remain a member in good standing" (Stamps, 2000, p. 60). Unlike other corporate units such as project teams and departments defined by the formal organizational structure of the organization, COPs "organize themselves, meaning [that] they set their own agendas and establish their own leadership" (Wenger & Synder, 2000, p. 142), forming around a "value-adding something" (Stewart, 1996, p. 174) that tends to let people "know when and if they should join [and] if they have something to give and whether they are likely to take something away" (Wenger & Synder, 2000, p. 142). In fact, Wenger says that the absolute worst thing a company or manager could do is try to manage COPs. Thomas Steward, writing for *Fortune* magazine, echoes Wenger: "Managing them can kill them" (p. 174).

Within corporations identifying themselves as knowledge or learning organizations (Drucker, 1993), employees become knowledge workers who "constantly apply and add to" existing bodies of knowledge, knowledge that can no longer be understood simply in terms of "dissemination of data" (Gongla & Rizzuto, 2001, p. 842) and that add value to the overall enterprise of a corporation. For instance, corporations such as Xerox, BPAmoco, Royal Dutch Shell, IBM, Dow Chemical, and The World Bank have been discovering that communities of practice allow for a greater sharing of knowledge and transfer of best practices, fostering a sense of belonging, developing better problem-solving abilities, and encouraging innovation. These abilities provide a more effective method for learning professional skills and behavior.

James Paul Gee (2000), and other scholars who talk about the new work order such as Reich (1999), argue that COPs—fed by the speed, flexibility, and adaptability of new capitalism—are the "crucial node at which business, schools, and society are aligning and merging in the new capitalism" (p. 50). The learning theory behind COPs fits well the pace of the new work order that now values flattened hierarchies, distributed systems, information networking, and innovation

---

[2] The term *communities of practice* has enjoyed various definitions, some of which are very simplistic: "Informal groups of individuals who have similar work related activities and interests" (Lesser & Everest, 2001), "simple expansions of one-on-one knowledge-sharing" (Burk, 2000), or "distributed groups of people who share a concern, set of problems, mandate or sense of purpose" (Tomoye, 2004). Each of these definitions is useful only insofar as they begin the process of defining such a complex concept. I prefer to begin this discussion with Wenger's definition because it focuses on continued learning through negotiation.

(Brown & Duguid, 2000a; Gee, 2000; Gee, Lankshear, & Hall, 1996). Because "most of the knowledge in the community of practice is tacit" (Gee, 2000, p. 53), access to that knowledge requires mutual engagement in practice with other members who share the meanings generated and constructed from that practice. Tacit knowledge, says Gee, Lankshear, and Hull (1996), is just the kind of "knowledge in and of practice that allows workers to add value to the enterprise. . . . Tacit knowledge is not stateable, but learnable only through immersion in a community of practice" (p. 66). Attempting to codify tacit knowledge by making it explicit risks losing the value added by that knowledge. Tacit knowledge helps members "help each other to make sense of non-canonical or unexpected circumstances" (Teigland, 2000, p. 155).

Although there are no specific deliverables attached to the knowledge and practice in COPs, the value added is measurable in the stories members tell about practice (Wenger, 1998; Stewart, 1991). These stories function as "common law, as a usefully underconstrained means to interpret each new situation in the light of accumulated wisdom and constantly changing circumstances . . . they act as "repositories of accumulated wisdom (Brown & Duguid, 2000b, pp. 105-106). Wenger and Synder (2000) advise executives managing COPs to listen to "members' stories" because they can "clarify the complex relationships among activities, knowledge and performance" (p. 145).

WHAT ACADEMIC RESEARCH TELLS US

Although the concept of communities of practice has not been fully integrated into the mainstream of technical communication scholarship, some references to this theory of learning exist. Beginning in 1979, Carolyn Miller established groundbreaking opportunities for research into "how to belong to a community" or "enculturation" for both research and pedagogy (p. 617; see also Miller, 1984), followed by Thomas Miller's (1991) understanding of professional writing as social praxis, Dale Sullivan's (1990) concept of apprenticeship as foundational for pedagogy, and Berkenkotter and Huckin's (1993) understanding of genre as socially constructed. In a special issue of *Technical Communication* on knowledge-management principles and practices, Corey Wick (2000) notes how organizations use various types of communities, including COPs, as a means for diffusing tacit knowledge (p. 518). In his discussion about the relationship between industry and academia, Stephen Bernhardt (2002) suggests that a "useful and powerful way to improve relations" would be "to establish shared communities of practice involving frequent, active, project-based cooperation" (p. 85). In the introduction to *Reshaping Technical Communication*, Barbara Mirel and Rachel Spilka (2002) discuss technical communication and its role in various communities, including communities of practice. **As one of those shared spaces, the Society for Technical Communication recently revised its mission statement to include a reference to COPs: "Creating and supporting a forum for communities of practice in the profession of technical communication" (see *www.stc.org*).** These brief mentions cannot do justice to the complex theory of learning communities of practice support.

Closer to a community-of-practice model and the goal of meaningful learning, many scholars have supported client-based and service-learning projects and approaches in teaching technical communication using activity theory (Spinuzzi, 1996) and theories about situated learning (Freedman & Adam, 1996). The service-learning approach in particular has enjoyed a great deal of success by immersing students in the practices of a participating agency (see, e.g., Matthews & Zimmerman, 1999; Wickliff, 1997), an approach now with its own textbook (see Bowden & Scott, 2003). The service-learning pedagogy requires students to work with members from a specific community creating genres, thereby introducing them directly to the ways communities of practice create meaning through language. One difficulty with positioning client-based and service-learning approaches, however, is the limited opportunity for legitimate peripheral participation, the meaningful engagement in practice as a legitimate member of a community. As "pseudoconsultants," students are not actual newcomers to a community, and therefore are not directly invested in its identity- and meaning-making activities. No matter how closely students actually work with members of a community, their access to what is tacitly known is limited by their outsider status. Access to this tacit knowledge is essential if they are to gain the expertise necessary for full membership because expectations for performance are determined by the practices of a community *in* the constitution of practice (Wenger, 1998).

As one way to encourage an identity of participation (although this approach is not identified as such) some pedagogies attempt to simulate workplace contexts in the classroom. But from their case study of a financial analysis class, Freedman, Adam, and Smart (1994) learned that efforts to simulate workplace practices are limited by the institutional contexts of the academy: "none of this know-how will have been made available through simulations, no matter how realistically or elaborately staged" (p. 221). Citing Knoblauch, these authors note that "workplace practices are embedded in additional layers of social reality and cannot be understood or learned—apart from them" (p. 221). Indeed, in *Worlds Apart*, Patrick Dias, Aviva Freedman, Peter Medway, and Anthony Paré (1999) argue that these practices do not lend easily to replication because the "context constitutes the situation that defines the activity of writing," which in turn affects genre decisions (p. 17). The technical communication classroom is affected by its own ideological, political, and administrative constraints stemming from the institutional (academia) and disciplinary (English and science/engineering) contexts defining its writing activities, increasing the difficulty of simulating workplace activities. Like service-learning approaches, no matter how real a project, no matter how real an audience, the "real audience for the students [is] always the professor— in his role as *professor*" (Freedman, Adam, & Smart, 1994, p. 203). Consequently, these **workplace projects cannot offer students the legitimate peripheral participation status granted to actual newcomers because they are not actually seeking membership in that community and therefore do not have a stake in its enterprise.**

Participation in practice means participating in the sharing of stories about that practice. This participation includes both listening to and telling stories in ways that demonstrate knowledge of a community's practice. The telling of stories in a community of practice is an act of creating new meaning to the repertoire of practices (Wenger, 1998, pp. 202-203). New meaning is created through hermeneutic actions because telling a story (and arguably listening to) simultaneously occurs with the interpretation of the events it chronicles (Bruner, 1991). Because "community members socially [construct] their world through the narration of stories, turning coherent data into coherent information" (Teigland, 2000, p. 170), this construction is what makes knowledge accessible to newcomers. In their study of "working knowledge," (knowledge that "originates and is applied in the minds of knowers," p. 5) Davenport and Prusak (1998), argue that to capture what is tacitly known, we must value narrative because "research shows that knowledge is communicated most effectively through a convincing narrative" (Davenport & Prusak, 1998, p. 81).[3] **Indeed, Wenger (1998) argues that stories can make it easier to transport knowledge because they call "upon our imagination" (p. 203).** The role of imagination in participation is essential because "stories can transport our experience into the situations they relate and involve us in producing the meanings of those events as though we were participants" (p. 203).

COURSE DESIGN: ESTABLISHING A COMMUNITY OF PRACTICE IN THE CLASSROOM

The difficulty with creating and sustaining a communities-of-practice model within academic contexts, specifically the technical communication classroom, stems from some of the limits imposed by academic units. Quarters or semesters simply do not offer the time needed for a community of practice to fully develop a repertoire of practices or for members to define the competency requirements for membership, making it difficult to develop and evolve a repertoire of practices. Because practices evolve through mutual engagement, students are also limited by time, finding it difficult to develop ways of being—or an identity—during one group project or one semester. Especially limited is students' access to a COP's practice stories, which is vital for developing an identity. This development requires both access to the stories about practice and the criteria for evaluating the community's ways of working, expectations for belonging, and the rules for negotiating meaning, and, especially (or essentially), how to develop an identity of participation. To develop an identity of participation, members must engage in the practices of the community by making "proposals of meaning" that add value to the repertoire of practices and that demonstrate "ownership of these proposals" through competent interpretations and articulation of new and existing ideas (Wenger, 1998, pp. 200-202). In this process, members continually negotiate this identity of participation, demonstrating a sense of control over the meanings they create (Wenger, 1998, pp. 134-137). **To sustain an identity of participation, members add their own stories to enrich the community's narrative accrual.**

---

[3] Education psychologist Jerome Bruner (1991) argues that to be successful, a story must provide an "intuitively convincing account," that is, it must indicate both what happened and why a story is worth telling. What makes it worth telling is its breach from a canonical story.

To provide students with opportunities for adding value in the introductory technical communication classroom, I require students to read a novel, which then becomes the context—or the narrative accrual—for all assignments (both individual and collaborative). The novel provides both the instructor and the students with a comprehensive narrative of which they both have full knowledge, access to the community's ways of working, expectations for belonging, and the rules for negotiating meaning. The novel represents the community's domain, or body of knowledge, and its repertoire of practices— information that students will need in order to mutually engage in practice and develop an identity of participation. I have used Frank Herbert's (1987) *Dune* and Scott Russell Sanders *Terrarium* (1985) because both novels work well to establish a context for action and, as fantasy novels, both stories depict entirely new worlds complete with established histories, characteristics, and discourses. As novels, they provide comprehensive and complete narrative accruals, and because narrative naturally encourages interpretative activities,[4] they also provide as close a replication to a community of practice as is allowable in the space of the classroom. As a context, both *Terrarium* and *Dune* provide students with a repertoire of practices necessary for students to learn about and understand competence requirements for practice as well as the narrative accrual necessary for developing an identity of participation, which further realizes the outcome for this pedagogy: students will be able to communicate effectively within and among communities of practice, demonstrating understanding of how these communities work, how they establish expectations for membership, and how meaning is negotiated.

ASSIGNMENT SEQUENCE: PARTICIPATING IN COMMUNITIES OF PRACTICE

This assignment sequence is designed to help students enter a community of practice—moving from newcomer to experienced member. By the time they have worked through these processes of membership, they should be able to add value (create proposals of meanings) and take responsibility for the creation of those meanings, producing primary and ancillary documents. They should also be able to assess their own performance in the creation of those documents. This four-part sequence, described in this section in the order outlined by the outcome, allows students to emulate as closely as possible, through the *Dune* or *Terrarium* contexts, the natural development of an identity of participation in a community of practice. All assignment scenarios are written within the context of the novel and take on an experienced member's voice in order to situate students within the boundaries of the novel. The instructor is always situated as an experienced member of the community addressed in the assignment. Appendix A includes a sample assignment sequence and Appendix B lists several sample assignments that create a rhetorical purpose for

---

[4] See Bridgeford (2004) for an extended discussion of the interpretative properties of narrative and its role in technical-communication courses.

reports as they relate to the communities-of-practice outcome and traditional technical communication skills.

Assessment and Evaluation of Documents

Two kinds of assessment occur in conjunction with this pedagogy: document evaluation and performance evaluation. Each document is holistically scored, using a method demonstrated for students during assessment workshops throughout the term. Document refers to the primary document created and its letter of transmittal using traditional characteristics of those genres. This document grade, however, is not the grade students receive for the project. Instead, students receive a performance grade, which is essentially an assessment of their ability to develop an identity of participation. Based on students' ancillary documents (i.e., individual activity log, self-assessment, evaluations of group members) and taking into account evaluation comments from the students' group members and the group activity log and assessment, I determine a performance grade for each student. The ancillary documents represent students' efforts to take ownership of the meanings they created in this project.[5]

Understanding Communities of Practice

The first assignment in the sequence is designed to introduce students to a community of practice by asking them to provide information about themselves in ways that mimic actual newcomer situations. Newcomers often spend initial meetings, in person or in writing, describing their own experiences (such as with a résumé and interview) in an effort to communicate how their knowledge fits with the community they hope to join. Experienced members of this community tell initiation stories that help them determine whether to grant the newcomer legitimate status and therefore begin learning how to develop an identity of participation. The assignment should request easily accessible information from the student such as previous experience with technical documentation in their particular disciplinary field.

Because this assignment is due early in the term, students have neither fully learned about the communities-of-practice concept nor have they finished reading the novel. But just as in real communities of practice, students are expected to begin engaging in the practices of a community before they have fully learned the expectations for competence.

Demonstrating How Communities of Practice Work

The second assignment in the sequence is designed to help students identify the particular characteristics of a community of practice. As newcomers become more enculturated within a community, they spend much of their time learning about the community through the stories experienced members tell to guide newcomers' participation in practices. Although still enjoying newcomer

---

[5] Evaluation forms used in my courses are available at www.tracybridgeford.com/resources/

status, students demonstrate their ability to try on an identity of participation by providing accurate and complete information about the community through a documentation activity, such as creating a fact sheet or an almanac entry. I deliberately assign shorter documents at this stage in order to avoid overwhelming students and to emulate boundaries established in actual communities of practice. Assignments should require students to identify, describe, and analyze the particular characteristics and their meanings within the fictional community of practice suggested by the novel.

Because this assignment occurs after students have completed reading the novel, they will have acquired the same level of access to a community's narrative accrual other students and the instructor have. At this stage in actual communities of practice, newcomers will often spend much of their time listening to stories about practice and reiterating those stories in order to demonstrate their understanding of the repertoire of practices within which they are participating.

Identifying Expectations for Membership in Communities of Practice

The third assignment in this sequence is designed to enable students to meet the expectations of full membership. This level of participation represents members' efforts to fully participate in a community's specific practice, signifying to experienced members their understanding of performance expectations, or what competent membership looks like in that particular community. Adherence to this social network involves identification of the rules for competent membership *in the act of* participation within the boundaries of those rules. Assignments situate students within complex circumstances, requiring interpretative activity that leads to concrete action, making recommendations or proposing new ways of thinking about practice for a community. I generally assign more complex documents that require extensive research, such as recommendation reports or proposals.

At this stage of developing membership in a fictionally based COP, students should be able to articulate the meaning of practice stories without prompting *in the act of* participation. The research component enables students to understand the processes members use to reinforce existing and incorporating new knowledge and practices into their ways of being.

Learning about Rules for Negotiating Meaning

The fourth assignment in this sequence is designed to enable students to add value to the community by demonstrating an ability to negotiate meaning. At this point, members generally work independently, trusted to preserve and contribute to the community's enterprise. Within the context of *Terrarium*, students negotiate meaning by interpreting existing practices in interesting ways or by adding to the story in ways that take into consideration the original story. This process of negotiation represents how well students can take what they know tacitly and make that knowledge explicit to an internal or external audience. I usually assign instruction sets and procedures or the creation of forms for this last collaborative assignment because explaining how things work affords students the opportunity to take ownership of the meanings they construct in concert with others. Both *Dune* and *Terrarium* depict unfamiliar conceptual and mechanical processes and objects, which empowers students to create meaning.

Because this assignment represents the penultimate project in this sequence, students should be able to articulate the role of the meanings they construct to the community's enterprise and repertoire of practices. The design component of this assignment encourages students to create meanings through images, text, and symbols.

SELF-ASSESSMENT: SUSTAINING MEMBERSHIP

Finally, students must be able to articulate the value of their contributions to the community. This self-assessment activity is akin to annual performance reviews in the workplace that often require workers to articulate their value to the community and its enterprise. But more important, this self-assessment is indicative of the role reflection plays in actual communities of practice. This final reflective stage affords students the opportunity to tell the story of their performance in ways that demonstrate not only their understanding of a community's ways of working, expectations for belonging, and the rules for negotiating meaning but also their ability to add value to that understanding. This assignment is, essentially, students' opportunity to tell the story of their own performance and to demonstrate why that story is worth telling.

CONCLUSION

By using a novel as a context for participation, students can enter a community of practice imaginatively, trying on the newcomer role with actual legitimate peripheral participation status. The novel provides a safe environment for practicing the communication skills necessary for success in the new work order as well as for engaging in mutual participation in an enterprise. Before entering actual communities, students learn, imaginatively, how to participate in the practices of a community and to recognize how tacit knowledge is shared and interpreted among members. They learn to recognize themselves in practice.

APPENDIX A
Example Assignment Sequence

ASSIGNMENT 1:
Understanding Communities of Practice

Individual Assignment: Explaining Previous Experience

Audience

District Manager Zuni Franklin

Situation

You've been hired temporarily by the Institute of Global Design in your particular area of expertise. In order to place you appropriately within the organization, the district manager requests a memo describing your previous experience working with technical documents and communicating and producing technical information.

ASSIGNMENT 2:
Understanding How Communities of Practice Work

Collaborative Assignment: Understanding Communities of Practice

Audience

Readers of the *Enclosure Almanac*

Situation

In order to accurately record the events leading up to the Ingathers' escape from Oregon City, the Preservation Institute has begun collecting information about the various communities associated with enclosure cities. As an assistant researcher at the Institute, you've been asked by the Research Director to create a one-page fact sheet about past or current communities from or connected to Oregon City for the soon-to-be published *Enclosure Almanac.*

ASSIGNMENT 3:
Identifying Expectations for Membership in
Communities of Practice

Collaborative Assignment: Reporting Information

Audience

Human Resources Executives at the Institute of Global Design (as represented by Dr. Gregory Passio and Dr. Zuni Franklin)

Situation

The year is 2031 and people have been living in the Enclosures for about five years. You've been working at the Institute of Global Design at Oregon City since moving inside the Oregon City enclosure. The Human Resources executives are concerned about the small number of applicants for positions like yours. Over the last 5 years, the applicant pool for your specialty has decreased by 50%. The executives are concerned that their recruiting efforts are not adequately drawing qualified applicants. They are especially concerned that their application procedures do not address the unique aspects of the application process in your specialized discipline; therefore, the applicant pool is low. When writing your report, you will need to "invent" reasons for this low applicant pool based on what you know about your particular field. This invention must be plausible within the context of *Terrarium.*

Drs. Passio and Franklin have asked you to write a report about the unique application processes in your discipline. This report **is not** a description of the general processes of applying for a job common to all disciplines (e.g., submitting a résumé, interviewing, etc.). The purpose of this report is to **identify, describe, and analyze** a *unique aspect of the application process for your particular discipline.* This report should adopt a critical (reflective) perspective about this process in terms of the unique attributes professionals in your discipline look for in an application candidate: technical knowledge, professional and personal characteristics, and expectations for performance.

ASSIGNMENT 4:
Learning How to Negotiate Meaning

Collaborative Assignment: Explaining How Things Work

Audience

New employees at the Institute of Global Design

Situation

You've been working at the Institute of Global Design at Oregon City for three months in your particular field. Dr. Franklin's administration has become concerned that new employees are not being trained sufficiently in a variety of procedures. Because you're new and have some perspective on this topic, she has asked if you would write up an instruction set for one of the technological and sociological processes associated with the enclosure. These documents will be used to orient new employees at the Institute who are not familiar with how things work.

SELF-ASSESSMENT: SUSTAINING MEMBERSHIP

Individual Assignment: Self-Assessment

Situation

You have been working at the Institute of Global Design for the past five months and the probationary period for your employment as a Class 1 Trainee is almost complete. Before you can be promoted to a Class 2 Associate, you are required to submit a written evaluation of your performance for this period. This performance will be reviewed by Dr. Zuni Franklin.

Complete the Performance Evaluation Form, attach your individual activity log, and submit your performance review to Dr. Franklin by **April 28**. Make sure that your comments fully support the rating you give yourself for each criterion. All narrative comments should contain rich descriptions of your work, both individual and collaborative, and point to specific examples from the last five months. Submit this form in a manila folder with all documents completed individually (e.g., activity logs, peer evaluations, self-assessment, etc.).

APPENDIX B
Creating Rhetorical Purpose for Reports

Aspects of Outcome	Example Assignments	Traditional Skills
Demonstrate understanding of how to participate in a community of practice	**Letter of Application:** You have applied for a position with the Institute for Global Design in response to one of the following position announcements listed below, which were advertised two weeks ago. So far, you have had one preliminary meeting with a Human Resources (HR) representative and feel confident that she liked you. Today you received a call from the HR Director, indicating that you are one of five other applicants competing for the job. The representative informs you that the second phase of interviews will be taking place in two weeks. The Human Resource representative tells you that the division supervisors have requested that you respond to the scenario below in writing prior to a second interview to determine if you will be called for a second interview. Using *Terrarium* as a context and what you know about the character addressed (division supervisors), send application materials indicating your continued interest in the position and, from the list below, identify, analyze, and evaluate three problems, the strategies used to resolve these problems, and which solution you think works best and why. You'll want to find some point of comparison around which your analysis will focus. Feel free to add to your interpretations of these problems as appropriate.	Audience analysis Document design Organization and order of Information Analyzing text

APPENDIX B (Cont'd.)

Aspects of Outcome	Example Assignments	Traditional Skills
	Almanac Entry or Fact Sheet: In order to accurately record the events leading up to the Ingathers' escape from Oregon City, the Preservation Institute has begun collecting information about the various communities associated with enclosure cities. As an assistant researcher at the Institute, you've been asked by the Research Director to create a one-page fact sheet about past or current communities from or connected to Oregon City for the soon-to-be published *Enclosure Almanac*.	
Demonstrate understanding of a community's ways of working	**Instruction Set:** You've been working at the Institute of Global Design (IGD) at Oregon City for three months. Last Monday you and a colleague volunteered to write up a process explanation and mechanical description or an instruction sheet as requested by Dr. Franklin's engineering team. These documents will be used to indoctrinate new employees at the Institute who are not familiar with how things work. They will appear as part of a larger employee manual that describes all facets of employment at IGD.	Audience analysis Document design Usability and readability Organization and order of Information Visual forms
	Forms: In 2026 the Enclosure Act was enacted to ensure protection of humanity from the inhospitable environment that Terra had become. This Act provided the necessary legislative activity for the construction and implementation of globe-like structures called enclosures as the new home for humanity. These enclosures have protected humanity from the chaos of the wilds for the last decade. Now, 25 years later, the Preservation Institute is assembling a historical collection of technical documents associated with the construction and implementation of the enclosure system for their new Enclosure Museum. In conducting this historical research, it was discovered that most of these early documents were destroyed by water damage before they could be preserved on CD-ROM. The Committee for Document Preservation has been working on creating replicas of these documents as part of their efforts to present a picture of the everyday practices of various communities (as represented in *Terrarium*). They are interested in how these communities worked together, how they communicated their expectations for belonging, how their values and beliefs play out in these everyday practices, and how they established rules for negotiating meaning.	

APPENDIX B (Cont'd.)

Aspects of Outcome	Example Assignments	Traditional Skills
Identify how members establish expectations for belonging	**Research Report:** The Historical Research Society (HRS) established by the Institute of Global Design has been charged with writing a research brief describing the historical precedents to the Enclosure Act. As one of its newest members, you've been asked to research defining moments in the history of environmental awareness in the United States. Namely, you've been asked by Gregory Passio to provide a summary and analysis of the Environmental Protection Act of 1970, which he plans to use to support his plans for Project Transcendence. Passio has also asked that you analyze these situations, indicating any connections you see between the EPA and Terrarium. In a memo addressed to Dr. Passio, summarize the main points of the EPA and the decision-making processes leading to the creation of The Enclosure in Terrarium.	Audience analysis Document design Locating and using information Planning and drafting
Learn how members agree on the rules for negotiating meaning	**Memo:** Phoenix Marshall is considering joining the Ingathers because he is fascinated by their approach to collaboration. But his knowledge about collaboration, the Ingathers' work practices, and other forms of collaboration is limited. Because you have worked with each group, he has come to you for information. Unfortunately, you will be out of town on the day he is free to meet. However, you agree to send him a synopsis of three different collaborative groups you've worked with, including Zuni and Gregory, the Ingathers, and Judith and her disciples. **Definitions:** Imagine that we have a hearing-impaired student in class who is accompanied by an interpreter. This interpreter's job is to translate what is said/going on in class to the student. It is not the interpreter's job to read class material. Because we can't expect this interpreter to read *Terrarium* in order to accurately translate class discussion, we need to create definitions and descriptions of terms specific to *Terrarium* to aid the interpreter. In pairs, choose one of the terms below and write formal (categorizes and characterizes), informal (differentiates), and operational (what it does or how it operates) definitions for that term. Once you have defined the term, each group should be prepared to share the definitions with the rest of the class.	Audience analysis Collaboration Accuracy Clarity

REFERENCES

Allen, J. (1989). The question isn't 'could' but 'should': The case against using fiction in the Introductory Technical Writing class. *The Technical Writing Teacher, 16*(3): 210-219.

Allen, J. (1999). Refining a social consciousness: Late 20th century influences, effects, and ongoing struggles in technical communication. In T. C. Kynell & M. G. Moran (Eds.), *Three keys to the past: The history of technical communication* (pp. 227-246). Stamford, CT: Ablex Publishing Corporation.

Berkenkotter, C., & Huckin, T. N. (1993). Rethinking genre from a sociocognitive perspective. *Written Communication, 10*(4), 475-509.

Bernhardt, S. (2002). Active-practice: Creating productive tension between academia and industry. In B. Mirel & R. Spilka (Eds.), *Reshaping technical communication: New directions and challenges for the 21st century* (pp. 81-90). Mahwah, NJ: Lawrence Erlbaum.

Bowdon, M., & Blake, S. J. (2003). *Service-learning in technical and professional communication*. New York: Longman.

Bridgeford, T. (2004). Storytime: Teaching technical communication as a narrative way of knowing. In T. Bridgeford, K. Kitalong, & D. Selfe (Eds.), *Innovative approaches to teaching technical communication* (pp. 111-132). Logan, UT: Utah State University Press.

Brown, J. S., & Duguid, P. (2000a). Organizational learning and communities of practice: Toward a unified view of working. Learning, and innovation. In E. Lesser, M. A. Fontaine, & J. A. Slusher (Eds.), *Knowledge and communities* (pp. 99-122). Boston: Butterworth Heinemann.

Brown, J. S., & Duguid, P. (2000b). *The social life of information*. Boston: Harvard Business School Press.

Bruner, J. (1991). The narrative construction of reality. *Critical Inquiry, 18,* 1-21.

Burk, M. (2000, May/June). Communities of practice. *Public Roads. U.S. Department of Transportation Federal Highway Administration*. Available online: http://www.tfhrc.gov/pubrds/mayjun00/commprac.htm. [Access date: December 27, 2004].

Burnett, R. (2005). *Technical communication* (6th ed.). Boston: Thomson Wadsworth.

Davenport, P., & Prusak, L. (1998). W*orking knowledge: How organizations manage what they know*. Boston, MA: Harvard Business School Press.

Dias, P., Freedman, A., Medway, P., & Paré, A. (1999). *Worlds apart: Acting and writing in academic and workplace contexts*. Mahwah, NJ: Lawrence Erlbaum Associates.

Dombrowki, P. M. (1994). *Humanistic aspects of technical communication*. New York: Baywood.

Drucker, P. F. (1993). *Post-capitalist society*. New York: HarperBusiness.

Freedman, A., & Adam, C. (1996). Learning to write professionally: "Situated learning" and the transition from university to professional discourse. *Journal of Business and Technical Communication, 10*(4), 395-426.

Freedman, A., Adam, C., & Smart, G. (1994). Wearing suits to class: Simulating genres and simulations as genre. *Written Communication, 11*(2), 193-225.

Gee, J. P. (2000). New people in new worlds: Networks, the new capitalism and schools. In B. Cope & M. Kalantzi (Eds.), *Multiliteracies: Literacy learning and the design of social futures* (pp. 43-60). London, England/New York: Routledge Taylor & Francis Group.

Gee, J. P., Lankshear, C., & Hull, B. (1996). *The new work order: Behind the language of the new capitalism*. Boulder, CO: Westview Press.

Gongla, P., & Rizzuto, C. R. (2001). Evolving communities of practice: IBM global services experience. *IBM Systems Journal, 40*(4), 842-862.

Herbert, F. (1987). *Dune.* New York: Ace Books. (Originally published in 1965.)

Karis, W. M. (1989). Using literature to focus attention: Rhetorical models and case studies. *The Technical Writing Teacher, 16*(3), 187-194.

Kilgore, D. (1981). *Moby-Dick:* A whale of a handbook for technical writing teachers. *Journal of Technical Writing and Communication, 11*(3), 209-216.

Lannon, J. (2003). *Technical Communication* (9th ed.). New York: Longman.

Lave, J., & Wenger, E. (1991). *Situated learning: Legitimate peripheral participation.* New York: Cambridge University Press.

Lesser, E., & Everest, K. (2004). Communities of practice: Making the most of intellectual capital. Available online: http://www-1.ibm.com/services/us/imc/pdf/g510-1622-00-esr-communities-of-practice.pdf [Accessed December 27, 2004]

Liebowitz, J., & Beckman, T. (1998). *Knowledge organization: What every manager should know.* Boca Raton, FL: St. Luci Press.

Matthews, C., & Zimmerman, B. B. (1999). Integrating service learning and technical communication: Benefits and challenges. *Technical Communication Quarterly, 8*(4), 383-404.

Miller, C. R. (1979). A humanistic rationale for technical writing. *College English, 40*(6), 610-617.

Miller, C. R. (1984). Genre as social action. *Quarterly Journal of Speech, 70,* 151-167.

Miller, T. (1991). Treating professional writing as social *praxis. Journal of Advanced Composition, 11*(1), 57-72.

Mirel, B., & Spilka, R. (Eds.). (2002). *Reshaping technical communication: New directions and challenges for the 21st century.* Mahwah, NJ: Lawrence Erlbaum.

Nelson, R. (1986). The use of literature in teaching technical writing. *The CEA Forum, 16*(2 & 3), 1-4.

Reich, R. B. (1992). *The work of nations: Preparing ourselves for 21st-century capitalism.* New York: Vintage Books

Sanders, S. R. (1985). *Terrarium.* Bloomington/Indianapolis: Indiana University Press.

Spinuzzi, C. (1996). Pseudotransactionality, activity theory, and professional writing instruction. *Technical Communication Quarterly, 5*(3), 295-308.

Stamps, D. (2000). Communities of practice: Learning is social. Training is irrelevant? In E. L. Lesser, M. A. Fontaine, & J. A. Slusher (Eds.), *Knowledge and communities* (pp. 53-64). Boston: Butterworth Heinemann. (Original publication: *Training Magazine,* February 1997.)

Stewart, T. A. (1996). Managing: Ideas and solutions. *Fortune, 134*(3). 173+.

Sullivan, D. (1990). Political-ethical implications of defining technical communication as a practice. *Journal of Advanced Composition, 10*(2), 375-386.

Teigland, R. (2000). Communities of practice at an Internet firm: Netovation vs. on-time performance. In E. L. Lesser, M. A. Fontaine, & J. A. Slusher (Eds.), *Knowledge and communities* (pp. 151-178). Boston: Butterworth Heinemann.

Tomoye Corporation. (2004). What are communities of practice. Available online: http://www.tomoye.com/whatwedo/whatarecops.htm. [Access date: December 27, 2004]

Wenger, E. (1998). *Communities of practice: Learning, meaning, and identity.* Cambridge, England: Cambridge University Press.

Wenger, E. C. (2000). Communities of practice: The key to knowledge strategy. In E. L. Lesser, M. A. Fontaine, & J. A. Slusher (Eds.), *Knowledge and communities* (pp. 3-20). Boston: Butterworth Heinemann.

Wenger, E. (2004). Communities of practice: A brief introduction. Available: http://www.ewenger.com/theory/communities_of_practice_intro.htm. Access date: December 27, 2004.

Wenger, E. C., & Snyder, W. M. (2000). Communities of practice: The organizational frontier. *Harvard Business Review*, reprint number R00110.

Wick, C. (2000). Knowledge management and leadership opportunities for technical communicators. *Technical Communication, 47*(4), 515-529.

Wickliff, G. A. (1997). Assessing the value of client-based group projects in an introductory technical communication course. *Journal of Business and Technical Communication, 11*(2), 170-191.

CHAPTER 10

Analyzing the Interactive Audience: Constructing a Communal Knowledge Base

Ann Kitalong-Will

> **OUTCOME:** Students will understand how audiences interact with businesses within new digital contexts, and how they construct communal knowledge bases within such spaces.

Over the past decade—driven, at least in part, by the increasing competitiveness of globalization—U.S. businesses have begun to expect a higher level of communication abilities in their employees. Not only must employees understand how to write, they must also be able to engage in effective communication practices within a complex web of relationships, often within digital contexts. In light of these changes, and with the increased use of digital communication technologies, the very nature of business communication has changed. "Business-to-business" and "Business-to-consumer" ("B2B" and "B2C," respectively) practices are commonplace, and "on-demand" responsiveness is expected. **In this new climate, clients and consumers not only *communicate* with businesses to purchase goods and services, but they *shape* business practices by contributing information to the knowledge bases that rest at the heart of organizational resources.**

Technical-communication instructors who hope to prepare students for this changing climate need to teach students to write effectively and interact with information in digital media environments, as well as to understand how supervisors, associates, and project team members use and respond to information in these environments. Students must be able to understand not only *how* but *why* businesses and their clients create, access, exchange, and interact with information in digital landscapes.

179

The assignment sequence in this chapter helps students practice their audience analysis skills and learn new ways of—and reasons for—interacting with information within digital contexts.

WHAT BUSINESS AND INDUSTRY TELL US

Communications Technologies and Business Practices

The past decade has shown great changes in business with the advent of the Internet and the advances in communication technologies. Communications technologies that developed in the 1980s through the 1990s (e-mail, chat, voicemail, cell phones, pagers, PDAs) have created a new landscape for businesses—one where B2B, B2C, and on-demand strategies are the modus operandi. Not only do businesses leverage the technologies, they *need* the technologies to compete in the global economy.

IBM refers to the new modes of doing business as *on-demand.* An on-demand business is one that learns to be "nimble, fast, and responsive"; and communication technologies play a key role in such transformations (Greif & Millen, 2003, p. I). Greif and Millen explain:

> The [on-demand] business will be able to exchange information smoothly across organizational boundaries. It will respond quickly to changing market conditions and form, revise, and reinvent relationships with external partners. It will innovate in its interaction with customers and will serve its own employee population more efficiently, translating savings into crisper focus on business goals and customer value (p. 1).

Responsive businesses recognize that a successful model focuses on customers: what customers want and how to deliver. In fact, customer value and satisfaction is now the most prevalent type of research that companies conduct (Garver, 2001, p. 41). In his 2001 article, "Listening to Customers," Michael Garver explains that while customer satisfaction has always been an integral part of business, it has now become one of the key measurements of business success. He writes

> Recognized as a key component to various business strategies, world-class companies now measure and manage customer value and satisfaction. In these strategies, **customer value and satisfaction is often a key performance measurement, a leading indicator of financial performance, an important diagnostic measure for continuous improvement, and a tool to manage competitive advantage (p. 41).**

Businesses now have many choices in how they communicate with their clients. B2B and B2C—two modes of conducting business over the Internet—are becoming common. (Consider Amazon, Yahoo!, or Google. Even brick-and-mortar businesses like Sears and Coca Cola are incorporating online models.)

As we move from a manufacturing economy to a more service-oriented economy, it is clear that **employees must have strong communication abilities: they must understand how to respond effectively to customers and business partners and how to help consumers *communicate* to organizations, to shape information and services in ways that meet their needs.** In his article, "The Relationship Imperative" (2003), Richard Edelman points out that **consumers are no longer passive, as has been the assumption of businesses in the near past. Instead, consumers "want to be part of the show" (p. 3). Because of the technological advances in communications, consumers have come to expect and even desire a high level of interaction with businesses.** Consumers tend to trust businesses that welcome their participation (p. 3). Edelman explains that, "[c]ompanies can no longer send discrete messages to distinct audiences, because today's dynamic marketplace sees information moving across audiences, often becoming scrambled or diminished" (p. 11).

In a study focusing on how consumers communicate and interact with companies through web sites, Pin Luarn and Hsin-Hui Lin (2003) point out:

> The online consumer gives time, cognition and effort to the experience of interacting with the web site, and gets an experience enabled by the web site that hopefully makes it easy to find needed/wanted products, to checkout quickly and to receive confirmation about all important aspects of the purchase, such as order-confirmation and delivery-tracking. In this regard, the product quality, service quality, and web site quality are also intertwined with each other (p. 159).

Luarn and Lin found that while traditional measures of customer loyalty were applicable to e-commerce (e.g. consumer satisfaction with the perceived value of their purchase), online consumers required more interaction with the web site to generate *trust* in the companies with which they deal. **Luarn and Lin argue that online customer satisfaction has less to do with price and more to do with the relationship the companies are able to build with their customers through communications and interactions within web sites (p. 163).**

In their discussion of trust relationships in e-commerce business models, Anthony Bryant and Barbara Colledge (2002) describe some strategies that e-commerce sites employ in order to build trust with their customers: online accounts to encourage repeat business, Web incentives for repeat customers, and the ability for valued clients to share information (p. 37). For instance, **increasing numbers of companies use their web sites to encourage customers and clients to interact with each other through online forums.** Among just a few examples of these practices are

• Online retailer Amazon.com provides a method for customers to write book reviews for other users to read when considering product purchases. Amazon provides a method for users to respond to reviews by rating the extent to which the review influenced their decision to purchase (or not).

- In 1997, IBM promoted a two-week chess match between International chess champion Garry Kasparov and IBM's computer, "Deep Blue." The promotion encouraged customer participation through various online conferences where individuals discussed chess and the Kasparov games. Over five million visitors participated in the online community to discuss and observe the chess match between Kasparov and Deep Blue (Boettcher, 1999).
- e-Bay, currently the largest online auction site on the Web, has built its business entirely around the interaction of its users and their exchange of information. Users post their items for sale, are rated on the quality of their service, purchase items from each other completely online. e-Bay users can, in fact, build entire businesses of their own around e-Bay's model of online interaction.
- During the 2004 presidential primaries, Democratic candidate Howard Dean created a fund-raising phenomenon with his Web presence. Through his Web site, DeanForAmerica.com, he raised over $800,000 in one day, which brought his funds to over $7 million (Weiss, 2005). His community of grassroots supporters was organized through his web site, which provided methods for making donations, exchanging information on issues, organizing locally, tracking Dean's movement on the campaign trail, and reading his blog.

Other businesses, importantly, have developed successful consumer interaction models that allow users to help build the knowledge bases around which the companies offer services. The attraction of such approaches is the communal knowledge that becomes available at these organizational sites. Again, some examples:

- Geeks to Go (www.geekstogo.com) is a community of technical support specialists who pool their knowledge about computer and software repair on a searchable online forum.
- Microsoft provides a means for users to interact online with company representatives or other users. At their support site, consumers can choose a "Self Support" option, or the "Assisted Support" option, working with a Microsoft employee on the problem by phone, chat, or e-mail. Microsoft also includes a community forums section where users can interact with each other.

In sum, **the users of online information—and the new forms of digital communication in which they are involved—are diverse, and their level of participation is high.** Effective technical-communications courses will address the changing face of the "audience" in these digital contexts. **Contemporary audiences are *not only consumers* of information, they *also shape the information* that organizations provide, contributing to companies' communicative projects by adding information to communally constructed knowledge bases.**

WHAT ACADEMIC RESEARCH TELLS US

Theory, Pedagogy, and Audience Analysis

In their article, "Redefining Communication Education for Engineers," Hirsch, Kelso, Shwom, Troy, and Walsh (2001) describe the role of communication in the engineering profession as "more than a matter of grammar and punctuation; [communication] is just as much a matter of creating effective graphs, holding productive meetings, understanding the needs of a client, communicating a proposed solution to that client, and writing a logical detailed plan that can be implemented" (p. 12). They describe an integrated approach to teaching engineering communication, one that draws on the expertise of both engineering and communications professionals.

Traditionally, technical-communication courses include an audience-analysis unit, often employing a conventional approach to understanding an audience. *Audience* is typically divided into categories: The Expert, The Technician, The Manager, and The Lay-reader. Each category includes certain characteristics. For instance, Managers are interested in the financial aspects of a project whereas Experts are interested in the theory. An examination of several introductory technical-communication textbooks (cf. Burnett, 2005; Markel, 2002) yields similar approaches to audience analysis methods:

- Audiences are divided into different categories.
- Some of these categories overlap.
- Once categorized, audiences are further described with demographic information, such as age, gender, education level, and income.

A key assumption is that once students understand the characteristics of their audience, they will then produce documents tailored specifically to that audience. The strength of this conventional approach is that it helps students understand how an *audience* informs document production. Within this framework, however, an audience is conceived as a passive entity, one that simply accepts the information and absorbs it. The approach can be misleading given new digital environments, because it implies that once writers know an audience, the documents produced will meet all the reader's needs.

As Edelman (2003) points out, however, audiences are no longer passive (p. 3) and knowledge is no longer simply passed along to end users. With the new communication technologies available to businesses, audiences expect to exchange information with companies, contribute to and interact with organizational knowledge bases and thus dynamically shape company services to meet their own needs.

In their article, "Communal Constructivist Theory," Marilyn Leask and Sarah Younie (2001) describe a pedagogical approach congruent with this new understanding of increasingly active audiences. They argue that knowledge is *communally* constructed and focuses on "the different forms of virtual and real community building and operation as well as the different ways in which knowledge is constructed, shared and reconstructed, published and republished by both teachers and learners alike" (p. 119). Leask and Younie's discussion of communal constructivism provides "a unifying theory that encapsulates the ways in which information and communications technology (ICT) enables learners to collaboratively create knowledge" (p. 117). It is a pedagogical approach to learning that allows for collaboration within and without learning communities. In particular, they examine online learning environments that support a communal constructivist pedagogy in international contexts. The European Schoolnet (EUN), for example, was created as a collaborative network of universities and organizations in 20 European countries, designed to "enhance the educational experiences of students across Europe" (p. 122). As an outcome to communal constructivist pedagogy, Leask and Younie point out that "[s]tudents' awareness that they are writing for an external audience is said to enhance motivation and add a real-world dimension to classwork" (p. 130).

Leask and Younie (2001) describe five aspects of communal constructivism in communication:

- Knowledge is built communally (rather than by the individual).
- Knowledge builders draw on actual situations to inform their knowledge building, rather than on theoretical situations.
- Information and communication technology support access to knowledge bases by others beyond the individual's zone of proximal development (as Vygotsky theorized in Leask & Younie, 2001).
- Technology allows individuals to create high-quality output of knowledge.
- Technology allows others to "build on, add to and republish this knowledge for their own purposes or in conjunction with the other creators of the knowledge" (p. 119).

Importantly, this pedagogy depends on communication technologies that facilitate audiences' interaction with digital knowledge bases. **Because technologies are changing the way audiences use information, it is important that communications courses address these changes and help students understand how interaction takes place, how knowledge is communally constructed, and what is at stake in connection with these changes.**

In his book, *The Virtual Community,* Howard Rheingold (1993) developed the concept of "online communities," which has greatly contributed to conceptualizing the technological constructs that drive user interaction online today. Rheingold describes the attraction to participating in online communities:

> My friends and I sometimes believe we are part of the future that [J. C. R.] Licklider dreamed about, and we often can attest to the truth of his prediction that "life will be happier for the on-line individual because the people with whom one interacts most strongly will be selected more by commonality of interests and goals than by accidents of proximity" (p. 1, html).

For Rheingold, online communities provide a way for users to build meaningful relationships with users worldwide. Rheingold describes a type of enthusiasm that participants in online communities demonstrate:

> You have to be careful with the addiction model as applied to the range of human behavior. Is a prodigy who practices day and night a violin addict? Perhaps. Is a great actor addicted to the attention of the audiences? Probably. Is addiction the proper lens for evaluating the violinist's or actor's behavior? Probably not. But nobody who has let her meal grow cold and her family grow concerned while she keeps typing furiously on a keyboard in full hot-blooded debate with a group of invisible people in faraway places can dismiss the dark side of online enthusiasm (p. 1, html).

As the owner of an Internet Service Provider, I can personally attest to the role of enthusiasm and consumer participation in the digital marketplace. Many of our customers participate in our company in various ways, through various online channels, and their levels of participation have increased over the past 7 years that we've been in operation. Some examples from our own business:

- One user developed a popular "Buffy the Vampire Slayer" community—a group of fans who meet regularly to discuss all things Buffy. They began as a chat group, developing into one that hosts a popular fan site, attracting many other Buffy fans around the world.
- Another user developed his own chess server hosted by our ISP, where chess hobbyists gather to play games with other chess enthusiasts.
- My eleven-year-old son frequents CheatCodes.com, an online collection of tips and "cheats" to video and computer games, a knowledge base built by video game enthusiasts.

Kimball and Rheingold (web site) describe the ways that online communities benefit companies, noting that "Search engines find facts. People provide solutions to problems. Networks of people can solve problems for each other. Online networks accelerate and globalize the process." As the above examples demonstrate, online communities have become knowledge-sharing devices, connecting people to solve problems and build knowledge bases. *Audiences* not only read information in these digital communities, they actively participate; and businesses have begun to respond.

In the assignment sequence, students examine the communally constructed web site Wikipedia.com, which provides an online environment that demonstrates how audience members interact with information and help to construct a community's knowledge base. Upon completion of the assignment sequence, **students will understand how contemporary audiences interact with information and help to construct communal knowledge bases that have changed business practices over the last decade.**

ASSIGNMENT SEQUENCE: WIKIPEDIA.COM— AUDIENCE ANALYSIS AND INTERACTION

Wikipedia is a free-content online encyclopedia that users are able to freely change. There are several levels of editors to ensure its authors adhere to editorial guidelines. From a communal constructivist perspective, it is truly an example of a "communally constructed" knowledge system. **Wikipedia represents a good example of how the new audiences interact with and shape information within digital contexts as discussed in this chapter.**

Although users participate in developing Wikipedia's content, there are guidelines that its participants must follow, not the least of which is to produce material in a responsible and consistent manner.

> Wikipedia is an encyclopedia written collaboratively by many of its readers. Lots of people are constantly improving Wikipedia, making thousands of changes an hour, all of which are recorded on the page history and the Recent Changes page. Nonsense and vandalism are usually removed quickly (Wikipedia: Introduction, 3/11/2005).

Wikipedia operates on what is referred to as the "copy-left" license, which is designed for free-content organizations. "Copy-left" stipulates that, while copying and modifying materials is acceptable, any materials produced as a result of this manipulation carry the same license. Copies may be sold, but must be available in a format that allows for further editing (see Michael Moore's chapter in this volume). Although Wikipedia content is not guaranteed due to its open-edit approach, it maintains a Style Guide that it suggests users follow.

Upon completion of the Wikipedia assignment sequence, students should

- understand **how many contemporary audiences interact with and help to build knowledge bases,**
- **understand the strengths and weaknesses of conventional audience-analysis methods when applied to digital environments,**
- **develop strategies for reaching audiences across different communication media,**
- **demonstrate an understanding of responsibility when developing written communication, and**
- **understand how "authorized" knowledge compares to "unauthorized" knowledge.**

Part I: Preview Wikipedia.com

The purpose of this first section of the assignment sequence is to help students become familiar with Wikipedia.com by exploring the site's encyclopedic documentation entries. The goal is to help them understand how both operators and participants shape knowledge within the site.

Students form teams with whom they work throughout the assignment sequence. The teams' first task is to explore Wikipedia.com by visiting the following entries:

- **Main page:** As the entry page to Wikipedia.com, it provides students with an introduction as both an informational source and as a community.
- **Help Page:** This section provides links to useful documentation, including a "how-to" section and a Style Guide.
- **Copy Left:** This entry provides users with the overarching philosophy behind Wikipedia.com. In addition to this entry, students review "Copyright," "GNU," and "Intellectual Property."

After teams explore Wikipedia.com, each team writes a technical description of the site (see Appendix A). The assignment asks students to

- include a general definition of Wikipedia.com, by describing some of the philosophical underpinnings of the web site, which students can find in the entries described above.
- discuss "Copyright" versus "Copy-left," and "GNU" and how they relate to Wikipedia.com.
- describe how Wikipedia.com is managed and who contributes to its content.

Part I of this assignment usually spans 1–2 class periods. Evaluation is based on **students' understanding of how the Wikipedia audience interacts with and helps to build this knowledge base.**

Part II: Apply Conventional Audience-Analysis Methods

Students begin by thinking about an audience in terms of conventional documents and then apply their analysis to Wikipedia.com. In their teams, students work through a traditional Audience Analysis to develop a preliminary description of both a primary and secondary audience (see Appendix A). They use these descriptions in Part III of the assignment as well.

After groups have written audience descriptions, a class discussion follows, typically discussing the additional information students need to know about Wikipedia audiences in order to write effective entries. Part II usually spans 2–3 class periods. Evaluation is based on students' **understanding of how the traditional audience analysis methods apply within digital environments, and what additional information is needed to understand the increasingly active role of audiences in shaping knowledge within those environments.**

Part III: Audience as Participants

As the Internet and other technologies become increasingly available (and needed) in business, it is becoming more common for projects to be produced communally or in teams. The purpose of Part III is to provide students with

firsthand experiences with the model of active audience participation—involving them as both an audience and a contributor in the Wikipedia effort. Through this activity, students should develop an increasingly refined understanding of how audiences interact with various online information bases.

Teams select an entry to revise at Wikipedia.com, printing the entry and discussing how to revise it. Students discuss (see Appendix A)

- what information/phraseology they like about the entry,
- what information/phraseology they find useful,
- what information is missing,
- how can the entry be improved,
- how they plan to revise.

Once students have discussed the above, they spend the next several class periods working on revision. Students are encouraged to revisit their chosen entry as often as possible. The content at Wikipedia.com is dynamic, and changes might be made to entries by other users as groups are revising.

Teams are required to follow Wikipedia's Style Guide and to read over the "How to" section, which provides directions to changing content.

After teams write 2–3 drafts, they turn in a final copy, along with the original version of the entry, and post their new version to Wikipedia.com (see Appendix A). Teams write a cover letter to turn in with their final revision, describing the strategies they used to reach different audiences, the style guidelines they followed, and any technological difficulties they experienced.

Part III typically spans 2 weeks as teams work through various drafts. Evaluation is based on the team's ability to **develop strategies for reaching different audiences and the responsibility they take in developing their texts.**

As students work in their teams to revise and ultimately post their draft, they build upon the descriptions of the Wikipedia audience developed in the previous phase of the assignment. They are evaluated on the process they use as a group to develop their entry, the final product, and their group's descriptions of their process (cover letter). This portion of the assignment sequence is a workshop in which class time is spent on team discussions, revisions, peer review, and team/instructor conferencing. By working through the different guidelines in the Style Guide, students develop an understanding of the responsibility expected of them when developing written communications for an audience.

Part IV: Follow-Up

Following the final posting, we spend some class time discussing how the documents were produced. In particular, we discuss how effective teams' final products were in light of the group's dynamic. We also discuss the following:

- How active audiences/consumers in digital contexts differ from more conventional audiences/consumers.

- The drawbacks for audiences that participate in communally constructing a knowledge base and what can be done to improve upon drawbacks.
- How reliable communally constructed information is and can be.

We also explore other communally constructed knowledge bases and compare them to information produced by organizations that take more conventional approaches to information production and consumption.

The following are some possible web sites that students can compare:

- **CheatCodes.com compared with Nintendo.com:** CheatCodes.com is a site that allows users to post "cheats" and tips to fellow video gamers, while Nintendo.com is the more conventional corporate site of the video game producer.
- **geekstogo.com compared with McAfee.com:** Both sites are dedicated to distributing information about computer viruses and how to inoculate computers. Geeks to Go is an independent site focusing on computer virus repair, while McAfee produces antivirus software.

To wrap up the unit, each team puts together a report that discusses how audiences both use and contribute to information and how these new activities shape organizations' relationships with customers/consumers.

Part IV usually spans about 1 week. Evaluation is based on student understanding of **how *authorized* knowledge compares to *unauthorized* knowledge and how audiences interact with businesses through knowledge bases situated within digital contexts.** Students discuss which forums they found most useful and what benefits and weaknesses exist in communally constructed knowledge bases where everyone learns and everyone educates.

CONCLUSION:
BUSINESS NEEDS AND AUDIENCE ANALYSIS

In a recent essay published in *Discover* magazine, author Steven Johnson (2005) describes the limitations of static information. He writes, "When publishers go out of their way to keep readers from interacting with digital text, they are literally stripping away its essence. The result is a crippled form that has no capacity for manipulation and little portability" (p. 23). The author of an e-book, Johnson's criticism of the publishing industry effectively illustrates what has become commonplace in business: texts that cannot be manipulated, revised, or changed by audiences are increasingly less valuable than texts that can. Johnson points out that, "reading electronic text is more of a dynamic process than reading traditional books—it's more participatory. **Manipulation is the *raison d'être* of digital text**" (p. 22). Usability studies suggest that digital text is more difficult to handle than print texts: digital text is difficult to read and rarely prints well. However, digital texts continue to gain in popularity because users can interact with them and shape them (p. 23).

In their article, "Career Paths of Five Technical Communicators," Loel Kim and Christine Tolley (2004) point out that audience awareness has become a key skill in job performance and success (p. 382). The ability to analyze audiences is an ability that they report as one of the "critical skills" needed on the job (p. 383). Organizations are reaching out to audiences who expect the ability to access, alter, and interact with the information that companies provide. The Wikipedia assignment sequence helps students understand both conventional and contemporary ways of understanding an audience and demonstrates how audiences have begun to interact with and add to digital knowledge bases.

APPENDIX A
Wikipedia.com: An Introduction to Audience Analysis and Interaction

---

ASSIGNMENT SEQUENCE DIRECTIONS

Part I: Explore Wikipedia.com

Purpose: The purpose of this assignment sequence is to help you understand how to apply some of the different audience analysis methods we have discussed in class and to explore the different ways that an audience interacts with information.

Directions: In groups, visit the web site, Wikipedia.com, and read over several of the entries. In particular, read over the following:

1. Main Page: This page serves as the entry point for the entire web site. It will provide you with an introduction to Wikipedia.com and introduce you to some of the community's practices.
2. Help Page: This section provides you with links to various Help documents available. In particular, take a look at the Style Guide and revision guidelines.
3. "Copy-left": Related to "copyright," this concept is a guiding principle at Wikipedia.com. You should also look up the entries for "copyright," "GNU," and "Intellectual Property."

Write a one-page technical description of Wikipedia.com. Consider some of the "philosophies" that guide users: How are "copy-left" and "copyright" related to Wikipedia? Who edits Wikipedia? Who writes for Wikipedia?

Part II: Analyze Wikipedia's Audience

Purpose: The purpose of this portion is to apply the audience analysis methods we have discussed in class to Wikipedia.

Directions: In your groups, identify primary and secondary audiences at Wikipedia. Fill in the Audience-Analysis Sheets to help you consider them.

Audience Analysis Sheet:

(Circle One): Primary Audience   Secondary Audience

Level of Expertise with Wikipedia:

Age Range: Professional Experience:

Where will your audience use Wikipedia?

What does the audience expect from Wikipedia?

How will the audience access Wikipedia?

What other characteristics should be considered?

Part III: Understanding the Audience at Wikipedia.com

Purpose: The purpose is to "put yourself in the audience's shoes" to see how audiences interact with "official" information produced by businesses.

Directions: In your groups, select an entry at Wikipedia.com to revise. Print the entry, and discuss the following:

- What information and phrases do you like about this entry?
- What information/phrases to you find useful?
- What information is missing?
- Is there any information that isn't necessary?
- Is there any part that your group finds confusing?
- How do you think the entry should be revised?

Be sure to record your group's answers to use later!

Once you have decided on revisions, write a new version of the entry to be posted to Wikipedia.com. You must follow Wikipedia's Style Guide. Make sure you revisit this entry often to view any changes that may be made during your group's revision process.

Print a final copy of your revision to turn in, and post the final copy of your revision to Wikipedia.com.

Part IV: Follow-Up

In your groups, visit some of the following web sites to see how audiences interact with official information:

- CheatCodes.com and Nintendo.com
- geekstogo.com and support.microsoft.com
- williamgibsonbooks.com and antonraubenweiss.com/gibson/index.html
- lordoftherings.net and zovakware.com/tests/lordoftherings.htm

What information is available at the "official" sites compared to the "unofficial" sites? How accurate and useful is the information at each site? Which forums did you find most useful? Why?

REFERENCES

Amazon.com. [web site] http://www.amazon.com. Accessed April 29, 2005.

Boettcher, S. [Web page]. (1999). Case Study: The IBM/Electric Minds *Kasparov v. Deep Blue Chess Match.* http://www.fullcirc.com/community/figalloibm.htm. Accessed April 29, 2005.

Bryant, A., & Colledge, B. (2002). Trust in electronic commerce business relationships. *Journal of Electronic Commerce Research, 3*(2), 32-39

Burnett, R. E. (2005). *Technical communication* (6th ed.). Boston, MA: Thomson Wadsworth.

CheatCodes.com. [web site] http://www.cheatcodes.com. Accessed March 22, 2005.

E-Bay. [web site] http://www.ebay.com. Accessed April 26, 2005.

Edelman, R. (2003-2004). The relationship imperative. *Journal of Integrated Marketing Communications,* 7-13.

Garver, M. (2001). Listening to customers. *Mid-American Journal of Business, 16* (2), 41-54.

Geeks to Go. [web site] http://www.geekstogo.com. Accessed April 2, 2005.

Greif, I., & Millen, D. R. (2003). *Communication trends and the on-demand organization: A Lotus workplace white paper.* Cambridge, MA: IBM–T. J. Watson Research Center.

Hirsch, P., Kelso, D., Shwom, B., Troy, J., & Walsh, J. (2001). Redefining communication education for engineers: How the NSF/VaNTH ERC is experimenting with a new approach. *Proceedings of the 2001 American Society for Engineering Education Annual Conference and Exposition,* Session 2261.

Howard Dean for America. [web site] http://archive.deanforamerica.com. Accessed April 28, 2005.

Johnson, S. (2005, May). Reinventing the e-book: If only publishers would let us cut, paste, forward, and even change their words. *Discover,* pp. 22-23.

Kimball, L., & Rheingold, H. [web site] http://www.rheingold.com/Associates/onlinenetworks.html. Accessed April 29, 2005.

Kim, L., & Tolley, C. (2004). Fitting academic programs to workplace marketability: Career paths of five technical communicators. *Technical Communications, 51*(3), 376-386.

Leask, M., & Younie, S. (2001). Communal constructivist theory: Information and communications technology pedagogy and internationalisation of the curriculum. *Journal of Information Technology for Teacher Education, 10*(1&2), 117-134.

Luarn, P., & Lin, H. (2003). A customer loyalty model of e-service context. *Journal of Electronic Commerce Research, 4*(4), 156-167.

Markel, M. (2002). *Technical communication* (6th ed.) Boston, MA: Bedford St. Martin's.

McAfee – Antivirus software and intrusion prevention solutions. [web site] at http://www.mcafee.com. Accessed April 12, 2005.

Rheingold, H. [online version]. *The virtual community: Homesteading on the electronic frontier.* http://www.rheingold.com/vc/book. Accessed April 28, 2005.

Weiss, T. Democrat Howard Dean strikes an online chord with campaign. [Web page] at http://www.computerworld.com. Accessed April 29, 2005.

Wikipedia, The free encyclopedia. [web site] http://www.wikipedia.com. Accessed March 11, 2005.

SECTION 3

Focus on the Complexities of Practice

CHAPTER 11

Understanding Usability Approaches

Johndan Johnson-Eilola and Stuart A. Selber

> **OUTCOME:** Students will develop a holistic understanding of usability, one that scaffolds common approaches according to their social complexities.

WHAT BUSINESS AND INDUSTRY TELL US

Usability is an interdisciplinary domain that is important to the work of technical professionals. Usability involves a set of practices that aim to improve the experiences people have with designed artifacts. Most historians cite World War II as the context within which usability became an organized endeavor. As Sedgwick (1993) explained, "The field has roots in Frederick Winslow Taylor's time-and-motion studies at the turn of the century, but it first flowered during the Second World War, when the fate of the Allies hinged on soldiers' ability to work complicated machinery" (p. 99). Since World War II, however, usability has expanded in three ways: Many different types of technical professionals, not just industrial designers and engineers, are now concerned with usability; many different types of artifacts, not just industrial and military machinery, can benefit from an attention to usability; and many different types of perspectives, not just those from scientific management disciplines, can contribute to an understanding of usability. In a very real sense, then, usability has insinuated itself into the mainstream practices of technical professionals, challenging them to be more attentive to the needs and tasks of real users in specific work settings.

Usability saves time and money. Economic and resource benefits are a major reason for the growing popularity of usability. Consider results from the following studies in business and industry. Nielsen (1993) reported on three organizations that saved significant amounts of time and money by focusing on usability (pp. 2-3). The first organization created a simple interface element that reduced demands on a centralized switchboard. This resulted in a total annual savings of around $1,000,000. The second organization redesigned application forms so that customers were less likely to make mistakes. This resulted in an annual savings of A$536,023 (Australian currency). The third organization increased the speed with which employees could log into a secure computer system. On the first day the system was in use, this resulted in a savings of $41,700. Likewise, Spencer and Yates (1995) noted that GE Information Services saved over $1,000,000 per year in support costs by providing customers with user-centered documentation. Karat (1993), a manager at IBM, discussed two usability engineering projects for business applications with impressive cost-benefit ratios. The cost-benefit ratio in the first project was 1:2 (IBM spent $20,700 on usability measures but saved $41,700 in overall costs), while in the second project the cost-benefit ratio was 1:100 (IBM spent $68,000 on usability measures but saved $6,800,000 in overall costs). In a discussion of how to calculate such cost-benefit ratios, Donahue (2001) presented a conservative scenario in which every dollar spent on usability returns $30.25 (p. 33). And in discussing the engineering approaches at Hewlett-Packard, Rideout and Lundell (1994) explained that usability is a key to both reducing costs and increasing customer productivity. These technical professionals emphasize an important point: usability can benefit both companies and customers alike.

Usability improves work processes. Savings in time and money are often a by-product of improved work processes. Although usability is frequently undertaken to increase user productivity, it invariably takes into consideration the ways companies function, which are not always as effective as they might be. For many organizations, implementing a usability program has resulted in more and better lines of internal communication, front-loaded design processes in which more and varied perspectives are listened to and considered, and development models that are more recursive and reflective. For example, Wroe (1994) made the case that usability standards should have a positive impact on each phase in the system development process. Latzina and Rummel (2003) demonstrated that corporate training sessions about usability can productively influence how technical professionals collaborate. And Bloomer and Croft (1997) argued that usability approaches can help companies achieve their business objectives. In each of these cases, usability was not simply added onto existing work practices. Rather, it played a more transformative role, improving the overall quality of those practices.

Usability increases user satisfaction. Companies are keenly aware of the fact that people today have an unprecedented number of options when it comes to selecting a product vendor or service provider. Although there are many different strategies for winning customers, usability moves beyond marketing products to making them as meaningful as possible to real users working in specific settings. The phrase "user satisfaction" obviously has many different meanings, but each one is valid and presents challenges to technical professionals. For Ong and Lai (2004), measuring user satisfaction involves paying attention to a wide range of software features, particularly those that enable users to customize system features. For Calisir and Calisir (2004), measuring user satisfaction involves paying attention to the perceived usefulness and learnability of computer systems. And for researchers like

Sikorski (2000), measuring user satisfaction involves paying attention to aesthetic and cultural issues. These technical professionals stress one of the greatest benefits of a focus on user satisfaction: it has legitimated the idea that human beings have affective relationships with designed artifacts (Norman, 2004).

WHAT ACADEMIC RESEARCH TELLS US

Rhetoric theory is not only congruent with usability perspectives, but extends them in useful ways. Teachers of technical communication have relied in central ways on both ancient and modern rhetoric theory. This theory has, for example, helped us to understand such fundamental areas as collaboration, style, document design, and ethics. It has even shaped definitions of technical communication itself. But rhetoric theory has also proven to be just as relevant to usability. Although technical professionals employ a different discourse, most approaches to usability could be considered to be rhetorical, for they at least partially focus on the audiences for designed artifacts, the purposes to which the artifacts will be put, and the contexts within which the artifacts will be used. For instance, Spool and his colleagues (1999) defined Web usability in purposeful terms: they are primarily concerned with what users hope to accomplish while online (pp. 2-3). And in her discussion of visual interface usability, Howlett (1996) said this: "If an interface is designed haphazardly, users won't know where to start or how to proceed. They won't know what's most important, and what is secondary, nor will they be able to find what they want in the mélange of visual stimuli. Completing their task will become slow and annoying" (p. 21). She, too, is interested in the tasks of users, and one piece of her advice is to design interface hierarchies that clearly distinguish primary from secondary functions. Nevertheless, usability perspectives have been extended by people working with rhetoric theory. In fact, our sense is that some of the most significant visions of usability are currently being advanced by the field of technical communication. Perhaps the greatest contribution is the idea that usability does not reside in an artifact itself. In their investigations, Spinuzzi (2003), Mirel (2003), as well as Grice and Hart-Davidson (2002) have all demonstrated that users of particular artifacts are governed not only by their built-in features but also by dynamic activity networks that include various actors, communities, texts, noncomputer tools, objectives, political realities, and more. Investigations like these trace the constellations of rhetorical forces that invariably influence the shape and appropriation of designed artifacts, promoting a view of usability that is distributed across time and space rather than localized in specific devices. Put in different terms, usability is evolving into an evaluative activity that is interested in the situatedness, politics, and social embedding of designed artifacts. Thus, rhetoric theory will become ever more vital to usability efforts.

The full spectrum of research methods in rhetoric and writing studies is crucial to understanding and measuring usability. When most people think of usability, they think of formal usability testing. With roots in experimental psychology, this approach takes place in laboratory settings that allow researchers to construct an environment that is conducive to observation and interaction.

Historically, the privileged lenses for examining usability have been experimental: studies are designed for controlled environments, variables defined, and results derived in quantitative terms. And there can be value in this type of empirical project, depending on the questions a researcher is asking. However, it is important to note that, given the generally pragmatic and fast-paced context of software development, researchers are more concerned with interpreting and understanding findings than with collecting data that offers a seemingly objective window into the usability of an artifact. For this reason, formal usability testing should not be unfairly stereotyped as the domain of old-fashioned experimental psychologists with social blinders. Rather, it is more accurate to think of it as one approach that can provide insights into the perspectives of real users and their tasks. But understanding and measuring usability requires more than a single method, as the previously cited investigations by Spinuzzi (2003), Mirel (2002), and Grice and Hart-Davidson (2002) make clear. These researchers combine the most productive aspects of the humanist critical tradition and social science methods, troubling the interpretive/empirical binary with multimodal usability approaches that address the complexity and chaos of actual situations of practice. Such approaches, for example, bring together ethnographic, participant/observer data gathering with close cultural analysis of texts, or they study social/cultural activity through a combination of qualitative and quantitative approaches. As Grice and Hart-Davidson put it, "Usability research today includes both the rigorous modeling of cognitive function and the thick ethnographic description of user behavior, both mathematical and narrative modes of analysis" (p. 161). The key is to permit the situation to dictate the approach, including unconventional combinations of method types. This perspective has been most cogently articulated by Sullivan and Porter (1997), who encourage researchers to think rhetorically as they design and conduct usability studies. Their view is that all methodology is rhetorical, involving assumptions, goals, techniques, and conventions forged in and across disciplinary contexts. This is not a flaw per se, but an inevitable aspect of the research process, one that should require people to argue for their methodological choices in each specific situation.

There are notable parallels between the histories of technical communication as a profession and usability as a practice. To reiterate, usability became particularly important in the Second World War, as soldiers struggled to learn and operate military equipment. Not surprisingly, this situation was also instrumental in the historical establishment of the field of technical communication. In fact, the history of technical communication reveals some additional factors that were influential in the development of usability. Souther (1989) argued that the field of technical communication was encouraged by three mandates for more effective public communication of scientific and technical matters: "(a) the environmental legislation, both federal and state, requiring that understandable environmental-impact statements be made available to the public, (b) the consumer movement that increased corporate and governmental responsibility for providing reliable information in plain language, and (c) the advent of the personal computer, which places the usability of software documentation at the center of the marketing strategy" (pp. 2-3). These mandates reflect several cultural developments that have driven usability work over the last half century or so, including the recognition that scientific and bureaucratic practices should be demystified within democratic contexts and the realization that both work and play are increasingly mediated through a range of digital environments. As in the case of technical communication, the mandates position usability somewhere between experts and users, attempting to make their relationship less hierarchical and more respectful, participatory, and human centered.

In our classroom practices, then, we draw on both what business and industry and academic research tell us about usability. We take seriously the concerns of the nonacademic workplace, asking future technical professionals to see themselves as usability workers who can help reduce costs, improve work processes, and increase user satisfaction. In doing so, we employ rhetoric theory and the full range of research methods in writing studies. We are also sensitive to the parallels that exist between the histories of technical communication and usability, which suggest the contributions that humanistic fields can make to the work of technical professionals.

ASSIGNMENT SEQUENCE: FOCUSING ON USABILITY

Although usability is a concern within any document-development context, the Web lends itself to useful discussions of user-artifact interactions in technical communication. There are several reasons why this is so. The Web draws attention to the reality that technical communication tends to get not so much read but used (Redish, 1993). People still read reports, proposals, instruction sets, and the rest, but not like novels or magazine articles. The field has long understood that readers of technical documents tend to read more opportunistically, employing texts strategically within the context of situated problem solving and work. This can be explained to students by way of the conventional discourse patterns that have emerged around printed technical documents to support purposeful reading, understanding, and use (e.g., headings, vertical lists, emphasis markers, given/new sequences). However, the Web includes certain features that further accentuate the reading/using distinction: hypertext links, which have both semantic and navigational qualities; search engines that support highly customized queries; and forms that turn texts into software programs that support various types of transactions and interactions. In addition, the Web stresses the multimodal nature of technical communication. If printed documents bring together words and images in complex ways, the Web advances this complexity by an order of magnitude, adding in the potential for not only audio and video discourses but also offering new methods for representing words and images. This situation provokes important discussions of usability in online documents, particularly because open-source databases have made it relatively easy to produce web sites with a rich array of multimodal features.

Finally, the Web diminishes the distance—both conceptual and physical—between technical communication and the products it supports. Technical communication on the Web is increasingly difficult to distinguish from the human/computer interface. Consider the users, goals, and time/space frames of three common forms of technical communication: tutorials, documentation, and help

(Selber, Johnson-Eilola, & Mehlenbacher, 1997). Tutorials are typically meant for novice users who have broad, long-term goals. The time/space frame for tutorials is all-encompassing: the tutorial is external to the environment it supports, and users typically give a tutorial their full attention. Documentation is typically meant for intermediate users with medium-ranged, short-term goals. In terms of time/space frames, documentation is used in parallel with the environment it supports: documentation shares the same work context, though not always the same system. These two forms of technical communication are relatively discrete, maintaining both conceptual and physical distance from the rest of the system. But help environments are another matter. Often meant for expert users with narrow, short-term goals, help environments are parasitic to the system itself, integrating tightly with the human/computer interface. Think of the keyboard shortcuts listed next to menu choices, the bubble help that appears when users move the mouse over an interface element, or the maps that help users navigate web sites. The usability of such items cannot be considered without assessing the overall structure of the human/computer interface. This convergence indicates an important direction in which technical communication has been headed over the last decade or so. Although there will always be a need for more loosely connected documents, the convergence of technical communication and interface design has the potential to influence the field in positive ways. It has certainly helped to integrate usability into the domain of technical communication, leading to new roles and responsibilities. For these reasons, the Web is a valuable context for pedagogical discussions of usability.

In the following section, we present a framework that teachers can use to introduce the landscape of usability. It is not meant to be a universal explanation, only one informative account of the dynamics and nature of usability work. The framework scaffolds three common approaches—heuristic evaluations, usability tests, and contextual inquiries—according to spatial aspects. The main advantages of this approach are that it foregrounds the social dimensions of usability and slowly discloses the complexity of usability. One danger is that students could mistake the scaffolding structure for a linear sequence. Students might assume that usability efforts should always begin with heuristic evaluations and end with contextual inquiries. This misconception can be avoided by stressing the inevitable recursiveness of usability evaluation.

Introducing the Framework

We were motivated to develop a framework because our students have struggled with usability. The basic idea is easy enough to demonstrate through everyday examples, but there are many different approaches and perspectives, which can be difficult to understand in a systematic manner, especially during one semester. In a previously developed scheme, we aligned a number of usability practices with typical phases in the document development process (Selber, Johnson-Eilola, & Mehlenbacher, 1997). For example, we roughly associated focus groups with early phases, protocol analyses with middle phases, and user/advocate reviews with late phases. This scheme is congruent with a process-oriented pedagogy, emphasizing recursion and the idea that usability should be both a formative and summative endeavor. We still use it with our students, who appreciate having a chronological structure that can be mapped onto the workflow of a project. Nevertheless, any one perspective is necessarily limited. A spatial approach represents another useful framework.

The framework involves metaphorical, rhetorical, methodological, and pedagogical locations. These cover both practice and theory with disciplinary sensitivity. We plot the locations in a matrix that includes three different types of usability that are increasingly social and spatial: heuristic evaluations, usability tests, and contextual inquiries (see Table 1). **Heuristic evaluations** are conducted by usability experts who analyze a digital environment against the best practices reported in the published literature. In heuristic evaluations, experts speak to one another about the environment, which is limited to the software itself. **Usability tests** open up this controlled feedback loop by asking real users to perform authentic tasks that can be observed and measured. Here the actions and reactions of the user/tester, which are psychological, emotional, and physical, constitute human layers of the environment. **Contextual inquiries** flesh out the human context because they jettison the controlled conditions of tests for the contingencies of user settings. **Contextual inquiries** attempt to understand a designed artifact as it gets employed in actual settings of work. The discussion below covers the spatial progression of these approaches in more detail.

Sequencing and Organizing Instruction

The assignment sequence is articulated to patiently disclose the complexity and diversity of usability practices. It requires increasingly broad encounters with the ecologies in which designed artifacts are both developed and used, challenging students to see usability as a social practice that involves various contexts, agents, forces, and approaches. The assignment sequence is thus informed by both the humanist-critical tradition and social-science methods.

Table 1. The Locations of Usability

LOCATIONS	Heuristic Evaluations	Usability Tests	Contextual Inquiries
Metaphorical	Usability as artifactual examination	Usability as experimentation	Usability as anthropological investigation
Rhetorical	Publication contexts	Writer/designer contexts	Reader/user contexts
Methodological	Analytical	Empirical	Multimodal
Pedagogical	Critical-analysis assignments	Peer-review activities	Service-learning projects

Before discussing the assignments, let us fit them into the larger instructional scene. **Like so many teachers of technical communication, we have evolved our courses to include more and more Web-based work.** Sometimes an assignment is added into an existing course that includes relatively traditional projects. Sometimes a more traditional project is revised to give it a Web-based component. And sometimes we dedicate an entire semester to the task of understanding the Web as a built environment for technical communication. But no matter the approach, the framework scales easily from the level of the assignment to the level of the course.

Because this last configuration is the one that structures our assignment sequence, we should elaborate on how usability can be implemented at the level of the full course. **The centerpiece of the course is a student-defined project with several associated documents.** We ask students to identify problems which have solutions that necessitate a new or significantly revised Web presence. Although the assignment is relatively open-ended, we do have a few parameters: the problems should be real and local, the audiences should be accessible, and the subject matter should be appropriate to present in a job interview. These parameters serve several purposes, including encouraging students to pursue their own professional interests and laying a foundation for the usability work. Students articulate the problem in a project proposal, which discusses the problem and its context, paying attention to rhetorical matters, resources, and constraints. The proposal also addresses outcomes, conventions, and initial design plans. Another associated document is a progress report, which in part functions as a milestone to keep students pushing forward on their projects. It updates the proposal, providing revised descriptions of the problem, audience, purpose, and so on; it also adds in node-link diagrams, design concepts, and discussions of progress on content development. As a final document, students write a commentary that reflects on their design choices. Meant to be more analytic than descriptive, students incorporate the course readings into a significant statement about their processes and products. This commentary encourages students to develop a metadiscourse for usability.

This brings us to the project itself. **Students are asked to develop two very different prototypes of the same web site, with multimodal and interactive features that are sensitive to rhetorical and conventional issues.** One task of the final commentary is to compare and contrast the prototypes—a task that reinforces the fact that technical-communication problems have multiple solutions, some of which may be better than others, but none of which is absolutely "correct." Because the question of scope is answered differently in each specific project, we consult closely with students to determine an appropriate workload. We also integrate the usability framework as an assignment sequence into the web site project. This is our focus here rather than the sequence of associated documents just discussed.

The first assignment in the sequence is a heuristic evaluation (see Appendix A for a sample assignment sheet). In this approach, students assess a web site using established usability principles. These principles, which reflect the findings of formal empirical studies, theoretical analyses, and practical work experiences, organize heuristics that can guide students as they inspect the usability of designed artifacts.

Heuristic evaluations are considered to be relatively quick and inexpensive. They do not require labs or special pieces of equipment, nor do they call for access to user sites. Moreover, heuristic evaluations are considered to be relatively easy to perform. In fact, as Table 1 illustrates, there are parallels between this method and certain practices already valued in writing studies. The spatial matrix constructs usability as an artifactual examination, a metaphor that emphasizes the critical analysis and evaluation of texts. The rhetorical location is the published research literature, a context that indicates the evolving knowledge base of an interdisciplinary field. This knowledge base gets deployed methodologically through text-analytic heuristic procedures that are alert to local social forces. To make the parallels clearer, consider the pedagogical location. In the service course, students are asked to critically analyze examples of technical communication. As part of the assignment, teachers provide students with a heuristic to help them generate ideas and responses. The heuristic derives from what the field knows about the rhetorical aspects of communication, but it has also been tailored to fit the situation. In groups, students produce a critique that attends to many different issues, including usability. Although not an exact match, there is considerable overlap between heuristic evaluations as an approach to usability and critical-analysis projects as an approach to technical-communication pedagogy.

There are four steps in a heuristic evaluation: planning the evaluation, selecting the participants, conducting the evaluation, and analyzing the results. Planning involves designing the evaluation instrument and identifying the heuristic. We start by thinking about an instrument our students can manage, one that is suitable to their usability experience. Beginners tend to benefit from structure, so for this audience students define a series of tasks that evaluators will be given to perform. In more advanced situations, we turn to scenario development, asking students to create realistic contexts out of which evaluators define the tasks for themselves (Rosson & Carroll, 2002). The least structured approach is appropriate to the most experienced evaluators, who are invited to "walk through" a prototype several times according to their own preference. In all three instances, however, heuristics provide the criteria for evaluation. The heuristics are typically taken from the established research literature and modified as needed to fit the situation. The pedagogical setting constrains the process of selecting participants. In an ideal situation, students would enlist three to five evaluators with expertise in both usability practices and the subject matter (Nielsen, 1993). As a rule, heuristic evaluations do not tend to uncover domain-specific problems, but it is difficult to evaluate form apart from content, which of course are mutually constitutive. Although students must recruit at least one ideal evaluator, the other people in the class also serve as evaluators, giving them organized experiences with both designing and performing heuristic evaluations. The performance part of the process begins with a meeting in which the evaluators are presented with an overview of the web site project and method of evaluation. They then conduct the actual assessment, identifying potential usability problems and tying them to specific principles in the heuristic. Finally, the evaluators are brought together to produce a composite list of the findings. This is a situation in which improvements to the prototype are inevitably debated. After the analysis session, the composite list is developed into a formal usability report. We will discuss the nature of usability reports in a later section.

The following example should make the process of running a heuristic evaluation more concrete. It does not touch on every facet, but instead illustrates some of the key moments. One of our students—we will call him Dave—created a web site for his brother, who is working as a film and television extra in Hollywood. Dave wanted to produce a site that would both help his brother locate work and educate aspiring actors about being an extra. Dave wrote a successful project proposal, which was due in the fourth week of class. Over the next several weeks, he made enough progress on the first prototype to begin the usability sequence. As to planning, the students in the course were new to heuristic evaluation, so Dave designed an evaluation instrument with a fair amount of structure. This consisted of a series of question tasks that were developed to assess the extent to which the prototype adhered to the usability principles catalogued by the heuristic. For example: What extra role did my brother play in the TV show *Malcolm in the Middle*? What professional goals has my brother set for himself in Hollywood? How many pieces of advice does my brother offer to aspiring actors? In conducting the evaluation, students worked recursively with the tasks and heuristic, staying alert to usability issues as they attempted to find answers to the questions. What sorts of issues did the evaluation identify? There were logical problems with both categories and headings (e.g., the advice for aspiring actors was found in several places); chunking and prioritizing problems (e.g., there were long lists that mixed film and television experiences and that subordinated the most impressive experiences); and navigation problems (e.g., the linking structure forced evaluators to rely too heavily on the navigation functions of their Web browsers). Ultimately, the composite list was formatted as a ratings form, which required the evaluators to rate the severity of the usability problems on a five-point scale. This is often the most difficult part of the process for students. In fact, one indicator of their educational progress is an ability to provide relative weightings and sort out major from minor usability issues.

The second assignment in the sequence is a usability test (see Appendix B for a sample assignment sheet). In usability tests, students work with real users and real tasks in order to assess a web site in development. This method collects empirical evidence that can steer and justify design choices. With roots in experimental psychology, usability tests often take place in lab settings that are conducive to observation and interaction. Although this approach is influenced by the tenets of experimental psychology, researchers are more concerned with interpreting and understanding findings than with collecting data that offers a seemingly objective window into the usability of an artifact. So usability testing has as much in common with the applied side of technical communication as with the disciplinary history of experimental psychology. There is also an overlap, as Table 1 shows, with certain pedagogical practices in technical communication. Like peer-review activities, usability testing offers ways for students to gain multiple perspectives on how their documents are read and understood by real readers (especially beyond the teacher, who many students see as a somewhat artificial and detached audience). Furthermore, unlike heuristic evaluations, which are centered primarily on the artifact itself and can call to mind certain current/traditional practices, usability testing situates artifacts in a slightly larger social space and, by bringing in representative users and tasks, expands that space in rhetorically useful ways.

The procedures for usability tests parallel those for contextual inquiries. Students must create a plan, select participants, run the test, and analyze the results. Usability testing, however, can get rather complicated at times, demanding expertise with a wide range of methodologies and software programs. Thus, it is important to discuss usability testing in ways that express its complexity without being overwhelming. This is not always an easy thing to do. In the planning phase, students define their goals and design the test itself. Defining goals is rhetorical in nature. Here, students address several questions: Who are the users? What are their tasks? What types of support should the web site provide? These might seem like straightforward questions, but answers typically are arrived at through a deliberative process of refinement and negotiation. Needless to say, a web site can support many different categories of users who engage in many different kinds of activities. So in attempting to set goals, it often becomes clear to students that they need to establish priorities, for no one empirical design can take into account even all of the basic issues. In the end, the audience and task definitions constitute a goals statement that clarifies the objectives of the usability test. Planning continues with the task of designing the test. The design process for empirical studies often begins with the formulation of a hypothesis. This is not the case with usability. Usability specialists are not typically interested in proving a hypothesis as either true or false. Instead, they are more concerned with transforming the goals statement into an apparatus that can test design features in meaningful ways with representative users and tasks. The testing apparatus is usually constrained by the artifact itself, a situation that highlights one recursive dimension of usability testing: Students should not select their method in advance, but allow the goals and the state of the artifact to help determine the method. Rubin (1994) distinguishes between exploratory tests, which target conceptual models and designer assumptions; assessment tests, which consider the effectiveness of specific efforts to create features that instantiate the models and assumptions; and validation tests, which examine how well a near-finished artifact compares to usability benchmarks. This scheme aids students in designing appropriate tests.

It is not easy to talk briefly about selecting participants, running tests, and analyzing results. These tasks differ according to the specifics of the usability test types, which are numerous and varied, to say the least. But an example from the classroom should help to clarify the full set of procedures (see Battleson, Booth, & Weintrop, 2001, for the details of a study on which the example was loosely modeled). One of our students—we will call this one Allison—created a new web site for a campus volleyball club. The proposed goals of the site were to advertise the club, promote communication among club members, and archive the results of club activities. Allison ran her usability test two weeks after the heuristic evaluation. She wanted to use the test to follow up on a navigation issue raised by the previous usability work. Allison employed a think-aloud protocol, a type of assessment test that asks students to verbalize their thoughts as they work with an artifact to accomplish real tasks. The design of the apparatus was not complicated. In fact, from a design standpoint, usability tests do not have to be overly rigorous. Students learn a great deal not by attempting to isolate and control the so-called variables of the situation, but by observing and listening to users. After much deliberation, Allison developed a set of navigation questions students would be asked to perform. The design issues involved determining questions that would yield sufficient data, wording the questions in ways that limited ambiguity, and scaffolding the questions so that they could build from each other. The protocol provided two students for each test, one to conduct the test and interact with the user, the other to record user movements through the web site and to take notes. The protocol also included a script to aid with consistency and support the note-taking process.

For each test question, the script included the ideal response, the least acceptable response, and instructions for when to intervene and end the test. It also created a space for a posttest evaluation, which asked users to further comment on the web site and on the test itself. In terms of conducting the test and evaluating the results, there are a few points to highlight. Allison took time to pilot the test, which yielded refinements to both the questions and the script. She also attempted to reduce bias by including several different students in the process. Multiple perspectives were also involved in evaluating the results. In drawing conclusions about the usability of the navigation system, Allison considered the think-aloud comments; the observations made by the note takers, including notes about body language, facial expressions, and reaction times; the observations made by the users; and posttest questionnaires. Also, Allison compared her findings to the usability principles on web site navigation used for the heuristic evaluation. This enabled her to situate the findings within a larger context, to see if the problems she noticed were also considered to be problems by the usability community.

The third assignment in the sequence is a contextual inquiry (see Appendix C for a sample assignment sheet). Contextual inquiry expands the focus of usability to include people employing web sites within actual working contexts. This method assumes that user contexts are much more complex than can be adequately understood in isolation. Contextual inquiry, then, moves outward from the artifact in isolation and from the usability lab out into the world. In addition, it typically includes aspects of ethnographic participant observation, with usability specialists sitting alongside users, asking the users themselves to illustrate their typical work habits, including things that are not directly located within the artifact being examined (e.g., interruptions, complex flows of social and technical information). The location of expertise within contextual inquiry similarly expands; usability experts are, to some extent, just along for the ride, with users themselves as the experts. So contextual inquires are typically more complex and more time intensive than heuristic evaluations and usability tests. Proponents point to greatly increased usability, particularly when usability is understood as a contextualized activity rather than an isolated act. As Table 1 suggests, the usability model described by contextual inquiry shares at least two points of connection with technical-communication pedagogy: face-to-face conferences and the use of real-world assignments, including service learning. Contextual inquiries are more akin to writing conferences than peer-review activities, offering students a more interactive and richer array of feedback opportunities, with writers able to ask questions in order to clarify interpretations and work out ideas in dialogue. And like service-learning projects, contextual inquires take place in user contexts, helping students gain the skills necessary to write for authentic situations. In short, this usability method demonstrates the contingency of meaning making within communities as a dynamic, interconnected activity.

Raven and Flanders (1996) offer a process for implementing contextual inquiries. While their focus is on software documentation, the general approach can be adapted to any type of artifact use in context. According to Raven and Flanders, contextual inquiries involve three principles: researchers gather data within contexts of use; researchers and users work as partners during the inquiry; and the inquiry begins by articulating one or more specific focus areas, rather than a list of interview questions or prespecified, functional tasks. These three principles guide the processes of contextual inquiry, which follow five coordinated steps: identify key sites for inquiry, identify key users at those sites, set the focus for visits to the sites, visit the sites and meet with specific users in their own workspaces, and analyze the information gained during the visits. Contextual inquiry should be a process that is relatively large grained and open to ongoing reassessment to ensure that the focus does not overlook issues that arise during the visits.

This approach can get complicated, so we require an initial assignment that provides practice with running contextual inquiries. It asks students to study the interactions between the architecture of a public space and the activities that happen within it. After selecting a public space for inquiry, students sit within that space. The spaces used vary wildly, often including such sites as a campus cafeteria and downtown coffeehouse. Students are prompted to take notes on the layout of spaces, the location of entrances and exits, seating arrangements, discussion and interaction patterns, lighting, and noise levels. Student observation notes reveal how these aspects support specific types of activities and resist contrary uses. Sometimes those uses seem "natural." Cafes, for example, support both fast access to a cook or server at the main counter while resisting (though not completely disallowing) customer-to-customer conversation, because customers are seated side by side and not face to face. In class discussion, students are also asked to comment on the variation from site to site in their studies. This activity is supported through a thought exercise in which students take the activities from one site (such as intense, individual work done in a dorm room) and place them into another site (such as an interactive, full-class workshop) in order to help them think about how both space and activity interact to structure each other.

Students move from the initial activity, which does not require them to actually interact with anyone, to the main assignment, which incorporates real users and their working contexts. The following is a small-scale example of a contextual inquiry from a geography student (John) who developed a web site for students taking Geography 101. The goal of the site was to help students with the major course project: a report that discusses the economic geography of a nonindustrialized country. John was an undergraduate teaching assistant for Geography 101, but as a student in the course, he struggled with several requirements, including writing progress memos, citing sources, and incorporating graphs and tables into written work. As a teaching assistant, he found himself answering questions about such matters repeatedly. So John decided to create a web site that would provide routine help on demand. He first identified three sites for inquiry: computer labs, dorm rooms, the library. These are primary spaces where students work on their reports. Next, he identified two users for each site, recruiting people from current sections of the course. John met with these users to set the focus for the visits. For all three sites, they decided to concentrate on the drafting phase of the report; based on previous experience, students tend to generate the bulk of their routine questions while drafting. The inquiry itself consisted of giving students an overview of the prototype and research methodology, watching them use the prototype while working on a draft, and discussing the prototype, draft work, and workspace. For example, John met Rachel at the library, which is where she likes to write. Rachel loaded the prototype web site on her laptop while John presented the overview. Rachel then proceeded to work on her draft for about an hour while John observed the entire process closely, taking notes, following her into the stacks, drawing maps of the space, and more. After the session, they studied the history trail of the Web browser and discussed how Rachel integrated the prototype into her drafting activities. John summarized the findings in a usability report that focused on how the space and activity structured each other, including the implications of that structuring for the design of the prototype.

The Deliverable

For each assignment in the sequence—heuristic evaluation, usability test, and contextual inquiry—students write a summative report that discusses methods, communicates findings, and reflects critically on the overall process. Students prepare these reports in addition to the other documents associated with the web site project: proposal, progress report, and critical design commentary. On one level, the deliverable could be seen as an iteratively designed prototype that incorporates the findings of considerable assessment work. But we are mainly concerned with helping students develop a holistic understanding of usability, one that scaffolds common approaches according to their social complexities. Consequently, we situate the summative usability reports as the deliverable for this assignment sequence, worrying less about Web-design execution and more about the ability to reason rhetorically and spatially about usability practices.

In terms of structure, the deliverables reflect two sources of knowledge about report writing: an industry standard from the usability community and guidance published in the literature on technical communication. In 2001, a common industry format for reporting on usability projects was approved as an ANSI standard (Scholtz & Morse, 2002). The standard echoes several laudable goals shared by usability specialists: advancing the status of the profession through written discourse, which demonstrates the value of usability perspectives, pro-viding concrete support for usability practitioners, and communicating effectively with clients. The scope and organization of the reporting format are relatively specific, advising writers, for example, to discuss the artifact being studied, the test objectives, the participants, the design of the test, and the results of the test. This structure reflects the typified organizational scheme for a scientific article: IMRAD (Introduction, Method, Results, and Discussion). Such schemes are undoubtedly useful to students, who need to develop an understanding of the disciplinary forces that shape discourse. However, the common industry format for reporting on usability assumes that people are running formal experiments. Indeed, one argument for the format was that it would make the experimental design of a study explicit to readers, allowing them to carry out replicating studies. So the format does not address how to report on heuristic evaluations or contextual inquires. This is why we turn to the guidance published in the technical-communication literature.

Rude (1995) admonished writers to match communication problems with appropriate research methods and genres. Her argument is that the organization of a report should not be determined in advance, for reports are neither static nor monolithic. Instead, writers should assume a rhetorical stance, organizing reports in ways that are sensitive to the specifics of the situation. As we have already mentioned, usability problems are ill-defined problems that involve a panoply of agents, settings, and goals. And the methods can be analytical or empirical or multimodal. In addition, the genre context can vary greatly, from highly informal internal reports to highly formal external reports. This writing scene is certainly complex, but it is also realistic. Corporate style guides for reporting on usability projects demonstrate such complexity: they tend to mix general conventions with local practices. In terms of general conventions, usability reports are similar to reports for decision making, which offer solutions to real problems situated in a specific time and place. We see usability problems as practical problems requiring action that is sensitive to social issues. The problems may indeed have theoretical or empirical dimensions, but the solutions are meant to aid decision making in working contexts. According to Rude, reports for decision making present findings and recommend solutions; they focus on the investigation itself and highlight criteria for evaluation. This is a useful articulation of several conventions for usability reports. Although the deliverable should reflect such conventions, it should also be responsive to local concerns. Let us quickly gloss a student example, to end on a more specific note about the shape of the deliverable.

Sarah and Dan were enlisted to run a heuristic evaluation for an instructional Web site on operating dendrochronology equipment, which foresters use to study tree rings and the growth patterns they represent. After conducting the assessment, their impulse was to organize the report in ways that mirrored the broad phases of this usability approach. This strategy obviously makes things easier on researchers, not readers. Sarah and Dan were therefore coached to be more rhetorically focused, taking into consideration the needs and interests of the audience members, who would be charged with either evaluating the usability recommendations in a managerial role or implementing them in a developer role. After reviewing sample usability reports and writing several drafts, Sarah and Dan produced a report with the following elements:

- Title page with contact information and date
- Table of contents with entries keyed to two levels of heading
- Executive summary with major findings and recommendations
- Introduction that discusses four areas: purposes of the web site project, goals of the usability procedure, methodological approach, assumptions of the researcher(s)
- Simplified findings organized in a table with two columns: Usability Issue and Severity Rating. The issues are grouped using the principles in the heuristic.
- Elaborated findings organized in paragraphs that integrate concrete examples of the problems, including screenshots. The paragraphs are grouped according to problem areas.
- Prioritized recommendations
- Appendices with the heuristic and rating scale

The report mobilizes a general-to-specific logic, progressively disclosing the details of the evaluation. It is also easy to use; the tables and headings support skimming and scanning. Moreover, the report foregrounds the goals of readers— receiving research-based recommendations for revising the prototype—and criteria for decision making, the usability principles in the heuristic. Sarah and Dan also accommodated local issues. For example, in the executive summary, they noted positive aspects of the prototype, being careful about their level of negativity toward a student designer who has proven to be sensitive to criticism. There are other features, but these are characteristic of a deliverable for heuristic evaluation.

Class Management

Managing the assignment sequence requires managing both usability work and the larger classroom context. There are management techniques specific to each usability approach, and teachers should become familiar with these before attempting the assignment sequence (see Kanter, 1994, for techniques for managing usability tests). Regarding the larger classroom context, there are just a few issues to consider—and these are rather unremarkable. One is collaboration. The usability sequence can overlay almost any type of assignment, from individual design work to group service-learning projects. But no matter the case, we like to put students into usability teams for the entire semester. The teams provide a constant source of feedback and support, modeling the collaborative nature of writing in the workplace. Another class management issue is timing. In our example of a course that focuses on the Web, the proposal is due during week four, the progress report during week eight, and the prototypes and critical-design commentary on the last day of class. Where does the usability sequence fit in? There are many possibilities. The main thing is to first afford students enough time to make progress on their prototypes. We often begin the usability sequence after the progress report is due. This report encourages students to make significant advances on the design plans discussed in the proposal by midway through the term. At that point, there are still plenty of weeks left for all three usability investigations. Ideally, there would be two weeks of (re)design time between each investigation, a pattern that stresses the iterative nature of Web development. The last issue is the role of the teacher. Everyone benefits when teachers position themselves as learners in the usability classroom. Teachers certainly have considerable expertise to share with students. Being student centered should not mean withholding or subordinating that knowledge. But usability is an evolving subject that draws on many different disciplines and methodologies. It demands an attitude of lifelong learning, especially when usability is understood to be an assessment practice that is situated in socially dynamic contexts. Teachers can model this attitude by including their own projects in the course. The course web site could become the focus of a warm-up activity that introduces the spatial framework. It should not be too difficult to imagine other ways to model an attitude of lifelong learning.

Assessment

The appendices include criteria for assessment for each assignment in the usability sequence. These criteria will be familiar to teachers. But more interesting matters arise from integrating usability into a course. As an assessment practice, usability is not very different from the grading teachers do. Teachers assess technical-communication products in rhetorical terms, provide concrete feedback for revision, and draw on disciplinary knowledge to determine what constitutes effective document structures. **The usability sequence, then, can serve a double function, helping to demystify the task of academic grading, as well.** As the social complexity of the assignment sequence unfolds, students experience firsthand the contingent nature of evaluating solutions to ill-defined problems. They come to see evaluation discourses not so much as truth claims to objectivity, but as informed comments shaped by research and local practice and the perspectives of evaluators. We extend this realization into the academic context whenever possible, involving students in the creation of course-grading criteria and in the evaluation of collaborative work.

Teachers with a rhetorical pedagogy can react to student questions in ways that seem to be inadequate. For example, saying "it depends" in response to a query about grading might very well be interpreted as a sign of indecision. Yet by the end of the course, students will be making similar sorts of conditional assessment statements. This will be a sign of their growing sophistication with usability.

CONCLUSION

This chapter offers a spatial framework for understanding usability. It can be integrated into any pedagogical context, serving as an overlay to both traditional and nontraditional assignments. It could also become the major focus of a course on usability. The framework responds to calls from industry for technical professionals who are prepared to contribute to usability endeavors. It is also informed by academic research that recognizes usability as a socially situated practice. The framework does not value one approach over another: heuristic evaluations, usability tests, and contextual inquires are all useful methods. Students, however, can have a difficult time trying to get a handle on the landscape of usability. The framework provides one way of thinking about the big picture. By discussing usability in spatial terms, teachers can provide an approach that foregrounds the social dimensions of usability and slowly discloses the complexity of usability.

APPENDIX A
Heuristic Evaluation Assignment

A heuristic evaluation involves four phases: planning the evaluation, selecting the participants, conducting the evaluation, and analyzing the results. It provides important feedback from reviewers who evaluate a web site using "best practices" garnered from research on human/computer interaction. Although heuristic evaluations do not provide feedback from actual users, selecting knowledgeable people to gather feedback makes heuristic evaluations an important component of usability.

After the heuristic evaluation, discuss your results in a report written to the designer(s) of the site.

1. Planning the Heuristic Evaluation

• Begin planning the heuristic evaluation by interviewing your client about what they expect users to do with the web site: Who will use the site? What should users get from the site? What representative tasks might users undertake?
• After interviewing the client, develop a list of three to five concrete tasks (in the form of questions) that your team of heuristic evaluators will use to guide the reviews of the web site. You should deal with concrete questions that users will be expected to answer when they visit the site, such as, "When is the campus gym open?" or "How can I apply for a grant?" Include questions that span a range of user activities in different parts of the site so that your evaluators can provide feedback on multiple areas and uses.

2. Selecting Participants

• Your heuristic evaluation team will need to include at least four people. Ideally, one person on the team will understand both web site design and the content area that the web site deals with. The other members of the team can be general Web users (including other students in the class). The heuristic evaluation team will meet twice: once to discuss the evaluation process and once to discuss the findings.

3. Conducting the Evaluation

• At the initial team meeting, introduce the evaluators to the web site, describing its purposes and users. If any team members are not familiar with basic Web design principles, provide a brief overview of relevant topics. Distribute the list of task questions you would like evaluators to use.

- Provide your questions in a form that lists task questions and provides prompts for responding in terms of basic usability principles, potential solutions, and general comments.
- After the evaluators have completed their work, gather them for a meeting. Ask each person to present their findings, following up each finding with discussion to analyze and debate the applicability of the finding to the web site and its users. These meetings can become argumentative, so you might need to mediate discussion. If necessary, remind the team that their meeting is only a part of the process, and the goal of the meeting is only to create a pool of possible findings that will be used later in writing the report.

4. Analyzing the Results

- Take the findings that the team has agreed on as well as disputed items, and draft a report to your client. Emphasize positive findings before addressing negative issues in order to avoid demoralizing the client (particularly if the evaluators found serious problems).
- Provide concrete solutions to issues that were identified. For example, if evaluators noted problems with terminology, offer alternative language that evaluators thought would work better.

CRITERIA FOR EVALUATION

- *Concrete feedback:* The report should provide concrete and focused feedback on the web site, including both summaries and specific quotes from evaluators.
- *Based on Accepted Design Principles:* The findings and recommendations should be supported by generally accepted site design principles.
- *Basic Report Structure, Format, and Style:* The report should reflect the standard structure, format, and style discussed in class.

APPENDIX B
Usability Test Assignment

Running a usability test involves four steps: creating a plan, selecting participants, running the test, and analyzing the results. A usability test can provide important feedback from real people completing realistic tasks. While a controlled usability test is a somewhat artificial situation, such tests are good for gathering information relatively quickly and focusing on representative tasks that the web site will support.

After conducting the usability test, discuss the results in a report written to the designer(s) of the web site.

1. Creating a Plan

- Think about the users of the web site. Your client will have some insight into the users, so include the client in your early thinking about usability testing. The client may also be able to provide you with some actual users of the site (whom you can contact for participation in the usability test). What sorts of people will use the site? Why are they going to be using it? What special characteristics will affect how they use and understand the site? Contact some actual or potential users of the site and gather information directly from them.

- After you have researched who might use the site, draft a description of those users that you can use in thinking about how to test the site.
- Draft a set of five to seven concrete tasks that users might typically engage in when they visit the site. Distribute the tasks among different structural and functional areas of the site, so that you will be able to get feedback that is applicable to different parts and functions of the site. For very large sites, select a subsection of the site, but your tasks should span different areas of that subsection. .
- After you have assembled a tentative list of tasks, walk through those tasks on the web site to see how long the test will take. You will not want to take up more than 30 to 45 minutes of your test subjects' time. If your walk-through takes substantially more or less time, revise your list of tasks accordingly.

2. Selecting Participants

- Select participants from the population that might actually use the site. If your client can provide names and contact information for such users, begin with them. If the site is aimed at a general audience, you may also recruit friends and classmates.
- Schedule at least five subjects to run the usability test. Although you can often gather usable results from a smaller number, five will allow a comfortable margin in case one subject misses their appointment or unexpected problems arise during the test.

3. Running the Usability Test

- Write up a list of specific task items that you can read verbatim to the test participants. Script an introduction in which you explain the test procedure to the participants, ensure them that you are testing the web site and not their own skills, and gather any necessary information about general Web experience as well as subject-area expertise.
- Construct a simple form that you can use to record the actions of the subject. Usually, a simple table with column headings for action, result, and notes will suffice. Even if you are videotaping the subject, this record will be useful in helping you analyze the results.
- Before you actually run the test, ask a classmate to act as a subject for a run-through of the full testing procedure. As the person works through the procedure, ask them to let you know if they are confused by any of your explanations or task descriptions. In addition, pay attention to the mechanics of the test to be sure everything works as expected.

- When each subject arrives, thank them for helping out with the test and explain the testing procedure. Give them an opportunity to ask questions.
- Ask the subjects to load the web site in their browser and provide them with a list of tasks. Ask them to verbalize their working and thinking processes while they complete the tasks. Take notes on your form as they complete tasks. If a subject runs into severe trouble, you may need to offer them advice or tell them they have helped you find a problem with a site and ask them to move on to the next task. (Note these interventions on your form.) Avoid providing too much direction to the subject, because this will skew the results of the test.
- After the participants have finished, ask them for general feedback about their experiences (which you should also log on your form). Then thank them for their participation.
- Review your own notes (and video recordings if they were taken), making sure that they will be comprehensible to you later when you write up your findings and recommendations.

4. Analyzing the Results

- After you have run all of the subjects through the test, read through (and view if applicable) the tests. What tasks could subjects complete without problems? What tasks did one or more subjects have difficulty with? Can you come up with recommendations that might alleviate those problems?
- Write up the results of your test and any recommendations in a report to the designer(s) of the web site. Describe the test procedure, including users who helped you test the site and specific tasks you asked them to complete. Provide rationales for both your selection of test subjects and why you chose the tasks you did. Provide test result summaries along with concrete details to back up the summaries (quotes from users, examples of specific problems, etc.). Conclude the report with concrete recommendations for improving the site.

CRITERIA FOR EVALUATION

- *Test Planning:* The usability test should include evidence of effective test planning, including subject selection and task list that targets a variety of sections of the site (or section of the site). (You should submit both the report itself and all additional materials—planning notes, test forms, videotapes, and so on—to the instructor.)

- *Test Results:* The report should outline the test procedure. In addition, it should summarize key findings as well as offer concrete examples to illustrate those findings. Any recommendations should respond to issues found during testing.

- *Basic Report Structure, Format, and Style:* The report should reflect the standard structure, format, and style discussed in class.

APPENDIX C
Contextual Inquiry Assignment

Contextual inquiry provides a method for gathering data about how people use web sites within the contexts where they will actually work. Although relatively labor intensive compared to other methods, contextual inquiry can provide a much richer body of feedback because the environments of users are often much more complex than you might expect, including aspects such as the state of their desktops, interruptions from co-workers, applications besides Web browsers open on their computer screen, and more. A contextual inquiry can reveal issues with a web site that might not emerge in heuristic evaluations or usability testing situations.

Although a full contextual inquiry typically involves multiple sites, you will need to conduct an inquiry at a single site. But be aware that single-site inquiries may limit the claims you can make about your findings.

After the contextual inquiry is completed, write a report on your findings to the designer(s) of the web site. In addition, provide a copy of the report as well as your notes to your instructor.

1. Identify Sites for Inquiry

• Discuss with your client who will actually use the web site, either existing users that the client is aware of or potential users.
• Ask your client to provide contact information about actual locations where the site is being used or might be used. Select one site for a possible contextual inquiry. If the client can provide contact information for a specific individual at that site, you will be able to use that information in the second step.

2. Identify Key Users at Sites

• If your client provided you with contact information for an individual at a site, contact that person directly to arrange a site visit. If the client was able to provide only a location but not a specific person to contact, investigate the location to determine a potential contact. Explain the purpose and structure of the contextual inquiry. Emphasize that the feedback will help make the web site more usable.
• Set up two meeting times with the participant: one to identify key issues for inquiry and one to actually observe the user at work with the site. Although contextual inquiries may sometimes take several hours, your inquiry will only last 30-60 minutes.

3. First Visit: Discuss the Focus of Inquiry

• During your first visit with the user, you will not actually observe them using the site. Instead, you will ask them a set of open-ended questions about how they might (or do) use the web site. If it is new to them, you will need to describe it and perhaps show them some screenshots.

- Work with the user to identify activities that they might complete on the site. When you schedule your second visit, try to schedule a time when the user might actually be using the site.
- After the meeting, review your notes while you look at the site. Locate areas of the site that the user talked specifically about or that correspond to activities they discussed. Identify any sections that correspond to problems or issues the subject identified. Note the location of these areas so that you will be able to observe all of them in your follow-up visit.

4. Second Visit: Observe the Users in Context

- On your second visit, begin by taking notes about the working context. Include details not only about the computer but also about the workspace. Is the area relatively peaceful, or are there constant distractions?
- Take one or more pictures of the site to help you recall the work context (and possibly use in your report).
- Ask the user to visit the web site and use it in a concrete way as they might normally use it. For an e-commerce site, for example, you might ask them if you can observe them placing an order for office products.
- As the user works, ask them to describe to you what they are doing. Ask for clarifications where necessary. After they have completed the activity, ask them to discuss how satisfied they were with the experience.
- Also ask the user to relate their experiences on the web site with their broader work context if they did not demonstrate that in their use of the site. For example, if they looked up the time and location for a meeting on the site, where did they record that information? If the information was copied and pasted from the browser into a computer-based calendar, did the text need to be edited before it was in a suitable form?

5. Analyze the Information Gained During Visits

- Review your notes from both the first and second visit. Did the site seem to match up with user needs and work contexts? Are there focus questions identified initially that did not seem to be addressed by the site? Were there any problems in using the site? How well did the site integrate with the broader work context of the user?
- Write up your analysis in a report for the client. Summarize your procedure and results gathered. Be sure to use information from your notes (and photographs, where relevant) to illustrate key findings. Make recommendations on possible site revisions.
- Give a copy of the report to your client and another copy, along with all of your notes and planning materials, to your instructor.

CRITERIA FOR EVALUATION

- *Inquiry Planning:* The planning notes should include evidence of useful discussions with the client about potential users of the site and summaries of key issues identified.

• *Site Visit Notes:* The notes from the site visit should be detailed and concrete, identifying key issues in both focus and inquiry visits. The inquiry visit notes should also provide in-depth documentation about the user at work, both in terms of their working context and the specific things they were doing with the site (user activities correlated with specific areas of the site visited).

• *Basic Report Structure, Format, and Style:* The report should reflect the standard structure, format, and style discussed in class.

REFERENCES

Battleson, B., Booth, A., & Weintrop, J. (2001). Usability testing of an academic library Web site: A case study. *The Journal of Academic Librarianship, 27*(3), 188-198.

Bloomer, S., & Croft, R. (1997). Pitching usability to your organization. *Interactions, 4*(6), 18-26.

Calisir, F., & Calisir, F. (2004). The relation of interface usability characteristics, perceived usefulness, and perceived ease of use to end-user satisfaction with Enterprise Resource Planning (ERP) systems. *Computers in Human Behavior, 20*, 505-515.

Donahue, G. (2001). Usability and the bottom line. *IEEE Software, 18*(1), 31-37.

Grice, R., & Hart-Davidson, B. (2002). Mapping the expanding landscape of usability: The case of distributed education. *Journal of Computer Documentation, 26*(4), 159-167.

Howlett, V. (1996). *Visual interface design for Windows.* New York: John Wiley & Sons.

Kanter, L. (1994). Techniques for managing a usability test. *IEEE Transactions on Professional Communication, 37*(3), 143-148,.

Karat, C. M. (1993). Usability engineering in dollars and cents. *IEEE Software, 10*(3), 88-89.

Latzina, M., & Rummel, B. (2003). Soft(ware) skills in context: Corporate usability training aiming at cross-disciplinary collaboration. *Proceedings of the IEEE 16th Conference on Software Education and Training* (pp. 52-57).

Mirel, B. (2003). *Interaction design for complex problem solving: Developing useful and usable software.* San Francisco, CA: Morgan Kaufmann.

Nielsen, J. (1993). *Usability engineering.* Boston, MA: AP Professional.

Norman, D. (2004). *Emotional design: Why we love (or hate) everyday things.* New York: Basic Books.

Ong, C. S., & Lai, J. Y. (2004). Developing an instrument for measuring user satisfaction with knowledge management systems. *IEEE Proceedings of the 37th Hawaii International Conference on System Sciences* (pp. 262-271).

Raven, M. E., & Flanders, A. (1996). Using contextual inquiry to learn about your audiences. *Journal of Computer Documentation, 20*(1), 1-13.

Redish, J. C. (1993). Understanding readers. In C. M. Barnum & S. Carliner (Eds.), *Techniques for technical communicators* (pp. 15-41). New York: Macmillan Publishing Company.

Rideout, T., & Lundell, J. (1994). Hewlett-Packard's usability engineering program. In M. E. Wiklund (Ed.), *Usability in Practice: How Companies Develop User-Friendly Products* (pp. 195-225). Boston, MA: AP Professional.

Rosson, M. B. & Carroll, J. M. (2002). *Usability Engineering: Scenario-based development of human-computer interaction.* San Francisco, CA: Academic Press.

Rubin, J. (1994). *Handbook of usability testing: How to design, plan, and conduct effective tests.* New York: John Wiley & Sons.

Rude, C. (1995). The report for decision making: genre and inquiry. *Journal of Business and Technical Communication, 9*(2), 170-205.

Scholtz, J. & Morse, E. (2002). A new usability standard and what it means to you. *SIGCHI Bulletin*, May/June, pp. 10-11.

Sedgwick, J. (1993, March). The complexity problem. *The Atlantic Monthly*, pp. 96-100.

Selber, S. A., Johnson-Eilola, J., & Mehlenbacher, B. M. (1997). Online support systems: Tutorials, documentation, and help. In A. B. Tucker (Ed.), *The computer science and engineering handbook* (pp. 1619-1643). Boca Raton, FL: CRC Press.

Sikorski, M. (2000). Beyond product usability: User satisfaction and quality management. *ACM Proceedings of the Annual Conference on Computer-Human Interaction* (pp. 61-63).

Spencer, C. J., & Yates, D. K. (1995). A good user's guide means fewer support calls and lower support costs. *Technical Communication, 42*(1), 52-55.

Spinuzzi, C. (2003). *Tracing genres through organizations: A sociocultural approach to information design.* Cambridge, MA: MIT Press.

Spool, J. M., Scanlon, T., Schroeder, W., Snyder, C., & DeAngelo, T. (1999). *web site usability: A designer's guide.* San Francisco, CA: Morgan Kaufmann Publishers.

Souther, J. W. (1989). Teaching technical writing: A retrospective appraisal. In B. E. Fearing & W. K. Sparrow (Eds.), *Technical writing: Theory and practice* (pp. 2-13). New York: Modern Language Association.

Sullivan, P., & Porter, J. E. (1997). *Opening spaces: Writing technologies and critical research practices.* Greenwich, CT: Ablex Publishing Company.

Wroe, B. (1994, December). The impact of standards on the systems development process. *IEEE Colloquium on Computer Safety* (pp. 6/1–6/3).

CHAPTER 12

An Ethics Primer: Strategies for Ethical Decision Making

Michael Martin

OUTCOME: Students will acquire a basic vocabulary and methodology for making ethical decisions in their technical communication practices.

WHAT BUSINESS AND INDUSTRY TELL US

After Charles O. Prince took the reins of the corporate giant Citigroup on October 1, 2003, he referred to the company as "serial sinner" in terms of ethical business practices. Prince once stated, "I never thought that you had to say to people, 'We want to grow aggressively—and don't forget not to break the law" (Schultz, 2004, p. 74). **The important connection between ethical concerns and corporate communication have recently become matters of debate as company scandals at World Com, AIG, and Enron have caught the public's attention.**

Ethical business communication is nothing new. Small businesses have long understood the importance of honest and effective communication. Some large companies—such as Gateway, for instance—built their reputation and pride themselves on promoting open communication within their corporate structure. Clearly, ethical business communication is important not only among individuals who interact directly and on a daily basis, but it also includes communicating honestly and frankly with individuals outside the business: clients, potential clients, and even competitors. **Ethical communication is also always relational. Even if the communication is electronic, the communicative process always occurs between two or more entities.**

The failure to communicate in ethical ways can yield both immediate and long-term consequences. Failure to do so often has had a ripple effect, altering business deals (Schultz, 2004, p. 72). Some information technology executives are now asking for proof of good ethical practices from vendors before agreeing to do business. Dell Computer Corp. has taken another approach to improving the ethical nature of corporate communications, appointing Thurmond Woodard as its chief ethics officer (Schultz, 2004, p. 72), a position unheard of a few years ago.

Effective communication and ethical corporate behavior are concerns for all companies. **Many small and midsize business owners state that they consider ethical business practices more important today than they have in the past** (Schultz, 2004, p. 72). In spite of that sentiment, however, a recent nationwide survey by George S. May (2005) revealed some disturbing news:

> . . . the owners' actions aren't matching their words when it comes to taking steps to improve ethical business operations; only slightly more than one third use any type of formal ethics training (para. 4).

The lack of any formal training in ethical communication creates a dilemma for both the individual and the company. **The lack of ethical training, especially in connection with the speaking and writing practices that occur within organizations, has specific consequences for faculty teaching introductory technical-communication classes. There is a critical need to provide students with an ethical foundation as they prepare for employment in a world where communication and business are intimate partners**. As communication networks increasingly extend a corporation's reach across linguistic, cultural, and geopolitical borders and as the public becomes aware of the unsavory practices of some business executives, the need for ethical training becomes ever more vital. Within such a context, effective communication becomes ethical communication, "truthful and relevant" communication that serves the interest of all the stakeholders (Bovée, Thill, & Schatzman, 2003, p. 20).

Ethical Decision Making, Trust, and Long-Term Consequences

Corporate dishonesty such as untruthful or unclear communication can result in the loss of jobs and the loss of personal retirement accounts. The loss of workers' retirement accounts, of course, is also a long-term consequence with exponential ramifications. George Brenkert, a professor at the Business Ethics Institute at Georgetown University, believes that as the number of corporate scandals grows, so do the consequences. He says,

> **There is a huge factor of trust that has been undermined by these scandals.** If there is no one to trust and if you don't think putting your money under your pillow is the best place, what do you do with it? There is a sense of powerlessness (Lauria, 2003, p. 3).

Enron's corporate collapse has also prompted a Congressional response. Both political parties have been called to account for the money Enron executives paid out in political contributions. A total of $5.7 million since 1989 bought a lot of influence in Washington, D.C. (Pellegrini, 2005, p. 1). In an appearance on *Face the Nation*, Senator John McCain (R-AZ) stated, "We're all tainted by the millions and millions of dollars that were contributed by Enron executives" (Pellegrini, 2005, p. 1).

More than 50 million Americans own stocks, usually in retirement accounts or managed funds (Lauria, 2003, p. 1). As people lose their retirement and savings, the executives of suspect corporations often walk with millions of dollars in stock options and salary (Hitt & Schlesinger, 2002, para. 8; Strauss, 2003, para. 5). WorldCom shareholders lost about $180 billion in WorldCom's collapse, 20,000 workers lost their jobs and the company went bankrupt (Ebbers Faces Prison, 2005, para. 12). The cumulative effect of lost retirement accounts and increased media coverage of a growing number of corporate fiascos has enraged the public. "It seems [they] simply can't take much more news of corporate scandals," said Diane Swanson, an ethics professor at Kansas State University (Lauria, 2003, p. 2).

Yet, there is good news; some business executives are working to improve the corporate image beginning with a change in corporate behavior. Nineteen global construction firms signed on to an "anti-corruption pledge" making a commitment to ethical business practices (Global Execs, 2004, p. 15). However, O. C. Ferrell, a professor at Colorado State University, predicts there are more scandals to come. The reason, according to business ethics professors, is a lack of ethical training (Lauria, 2003, p. 2). Jesuit Fr. Martin Calkins, a professor at Santa Clara University, notes with a sense of frustration:

> It is frustrating to teach business ethics. . . . We've been at this for 30 years as a scholarly group, and we just seem to be getting no where. There just seems to be one scandal after another (O'Neill, 2002, p. 1).

Kirk Hanson, a professional ethics consultant, has called for the establishment of regional centers to help both schools and businesses strengthen ethics training (O'Neill, 2002, p. 1). Yet, overall, as K-State ethics professor Diane Swanson says, "there is no real clamor among the deans of business schools to teach ethics"— it seems that several business schools have downsized the ethics coursework requirement (Lauria 2003, p. 2). Such a response, at least in the current climate, seems untenable.

Simon Webley and Martin Le Jeune (2005), two international ethicists, surveyed various groups using Codes of Conduct. Their survey revealed that "ethical codes are a mechanism" for establishing and articulating the corporate values, responsibilities, obligations, and ethical ambitions of an [organization] and the way it functions (Corporate Use, para. 3). But often codes are a litany of "what to do" or "what not to do." As a result, "reliance upon codes [becomes] the halfway point between visceral devotion to gut instincts and the application of ethical reflection and reasoning" (Steele & Black, 2000, para. 2). WorldCom CEO Bernard Ebbers' chief defense, for example, was that he didn't know what was happening in his own company. Therefore, he was not ethically responsible. The jury disagreed.

If corporate Codes of Ethics generally function in this manner, they are ineffective. **Simply put, the fact that an organization has a corporate Code of Ethics does not automatically translate into ethical behavior.** Nor do good intentions, as Stuart Gilman, the president of the Ethics Resource Center argues:

> The vast majority of businesses and corporations want to do the right thing. What Enron reminds us of is how vulnerable organizations are to ethical landmines (Espie, 2002, para. 1).

Effective Communication Fosters Ethical Behavior

If codes of conduct are to be effective, it is essential they are "actual living documents" (Espie, 2002, para. 3). "Codes of conduct are an outgrowth of company missions, visions, strategies and values" (Espie, 2002, para. 4). One difficulty is that Codes of Conduct rarely change once they are adopted. The Association of Educational Communication and Technology (AECT), for example, revised their Code of Conduct document in 2001—the first such revision in over 30 years (Yeaman, 2002, p. 12). There are, typically, few clear and dramatic calls or opportunities for revising organizational Codes of Conduct (2002, p. 11). The association realized the need to revise the organization's code to show more "social responsibility" (Yeaman, 2002, p. 12); they concluded that **"ethical issues are the responsibility of the [organization]"** (Yeaman, 2002, p. 12).

The nature of ethical communication is also complicated by both technology and the increasingly global marketplace. As companies have become more global, for instance, CEOs have increasingly turned to technology to communicate with rank and file workers. Such technologies, however, never convey information in the same way as face-to-face exchanges—especially in situations that are characterized by possible ethical conflicts. Understanding the ways in which technology, communication, and ethical behavior overlap are important aspects of ethical communication. As a Chrysler employee at the time of that company's merger with Daimler in November of 1999, I listened as the company's new CEO, Jürgen Schrempp, spoke of his vision for this new global company—on large screen televisions set up around the plant. Schrempp's address to the American work force—carried off with the help of new technologies—did little to assuage the belief that Daimler had just pulled off a nonhostile hostile takeover of one of America's Big Three automakers.

Similarly, when I worked for Gateway Computers, employees recalled fondly that Ted Waitt, the founder and former CEO, at the height of the company's success, regularly walked through the call floor to speak with employees. Globalization and rapid expansion outside the company's initial mission as a make-to-order computer company, however, had significant consequences on both the company's performance and their communication practices. In April 2004, Gateway was forced to close their Gateway Country Stores and lay off 2,500 workers (Gateway to close retail stores, 2004, para. 3). Corporate offices relocated to Southern California. In the wake of these changes, employees felt that leaders ignored some core values, including a lack of communication with upper management. In fact, Ted Waitt's personal move to California was interpreted as a change in communication and company focus. While I was an employee, attempts to stem this criticism came through a weekly e-mail from Gateway's new CFO. However, a weekly e-mail did not take the place of the pony-tailed Waitt strolling up and down the call floor.

Teaching Ethical Decision Making in Technical-Communication Classrooms is Vital

If unethical behavior is increasing and ethical training is decreasing, the need to connect effective communication and ethical behavior is growing exponentially, especially in technical-communication classrooms. As a result, instructors need to make time to speak to students about the intersection of ethics and communication.

Teaching students that the study of ethics is more than ideology, Kenneth Goodpaster of the University of St. Thomas argues that the study of ethics has "got to deal with the stuff of leadership and management" (Ferrell & Ferrell, 2002, p. 17). The question is how and when.

WHAT ACADEMIC RESEARCH TELLS US

Both scholars and practitioners agree that ethics is an important topic in technical and professional communication. Three well-known texts are *Ethics in Technical Communication: Shades of Gray* (Allen & Voss, 1997), *Ethics in Technical Communication* (Dombrowski, 2000), and *Ethics in Technical Communication: A Critique and Synthesis* (Markel, 2001). Often teachers of introductory courses in technical communication have a limited background in the formal study of ethics, limited time, and few strategies at their disposal to address such issues. This chapter offers an introductory ethics primer that provides some basic language and practical applications that teachers can use with students as they think, talk, and do ethics (Dombrowski, 2000, pp. 4-6).

The goal of such assignments is that both teachers and students come away with an enriched understanding of how ethical decision making is central to communication practices and specifically to technical communication.

Allen and Voss (1997) state "this book is written to help professional technical communicators and students . . . to deal with complex ethical issues, both on the job and in the classroom" (p. xix). They contend that the study of ethics is of "vital importance to the profession" (p. xix). **Ethics cannot be merely ideological, it must be applicable.**

Most textbooks that focus on ethics include critical reflection as well as practical application (Allen & Voss, 1997; Dombrowski, 2000; Markel, 2001). In respective articles, Carolyn Miller (2004) and David Russell (2004) argue that communication instructors must help students learn to hone their communicative skills, not just for the classroom, but as developing professionals (pp. 156, 166). Texas Tech professor Sam Dragga (2001) writes that "the information age . . . radically changes the practice of ethics [in the workplace]" (p. 247). Helping students navigate this change is one of the central tasks teachers have. Robert Johnson goes even further as he argues that ethical practice is an integral part of user-centered practice that highlights both the need for agency and good problem-solving skills (1998). Such an understanding connects the classroom teaching of technical communication to workplace communication. **If teachers are to address the tensions that exist between workplaces and individuals—to use "knowledge and skill for the common good" as James Dubinsky notes (2004a, p. 4), they need to provide students not only an ethical "toolkit," but also the ability to use the tools in their communicative practices (2004a, p. 5).**

In *Teaching Technical Communication: Critical Issues for the Classroom,* Dubinsky (2004a) argues that ethical action within the field of communication is not an abstract theory, it is a practical art (p. 7). It is also good rhetorical practice. With reference to Aristotle, he says that "praxis means being guided by a moral disposition to act truly and rightly. . ." (2004b, p. 245). Yet, rhetorician and scholar Carolyn Miller (1979) argues that it is *often* difficult to create an atmosphere where such concerns are foregrounded. She contends that technical-communication classes often focus on "the instrumental aspect of discourse—its potential for getting things done—and at the same time to invite a how-to, or handbook, method of instruction" (p. 612). Given this context, teachers within these classes should make sure that the writing tasks that students undertake in technical communication classrooms—manuals, online documentation, memos—are all discussed within the context of solving problems and making peoples' lives better. As David Russell (2004) notes, **technical communication is never simple or "value-neutral" (p. 166) rather, it involves multilayered processes and products that affect multiple audiences.**

Both Miller and Russell note another inherent tension in technical-writing courses. Subject matter experts, these authors explain, often guard their knowledge jealously. However, if writers want to be taken seriously within a field or become part of a community, they must also establish a professional *ethos*, and they must share the field's ethical values, rules, and ethical principles—explicit or understood (Russell, 2004, p. 167). **Ethical dilemmas often manifest themselves when professionals find themselves negotiating communications between fields and professional groups. Teachers can help student writers navigate these audiences and issues by providing them with effective strategies for making sound ethical decisions.**

ASSIGNMENT SEQUENCE:
A PRIMER FOR ETHICAL STRATEGIES

As instructors in an introductory class, it is important to prepare students for the ethical dilemmas they might face. The following sequence of assignments is intended to offer both a vocabulary and a method to effectively speak about and address ethical situations. As educators, it is vital that we connect their conversations with their experiences.

The assignments that follow offer a chance to read, reflect, and discuss ethical dilemmas in real life by providing

- an ethical vocabulary (Appendix A);
- a method for making ethical decisions (Appendix B);
- a case-study approach that allows students an additional opportunity to employ their shared vocabulary and methodology (Appendix C).

This series of assignments serves students' needs for considering ethical issues.

Establishing an Ethical Vocabulary

A conversation about the definition of ethics—and the terms used to discuss ethical decision making and communication—is a good place to begin in technical communication classrooms. Although the primer that follows provides some basic definitions of key terms, it is important to have students help define these terms within the context of their own experiences. These definitions, therefore, are offered only as a place to begin this exploration. They are listed alphabetically for simplicity.

Behavior: the manner of conducting oneself; anything involving action and response to stimulation

Code of Conduct: a code of ethics; a vehicle for communication among all the members of a group

Ethics: the discipline dealing with good and bad, moral duty and obligation; a set of moral principles or values

Ethos: the distinguishing character, sentiment, or moral nature of a person, group, or institution

Fairness: marked by impartiality and honesty; free from self-interest, prejudice, or favoritism

Justice: the quality of being just, impartial, or fair; the principle or ideal of just dealing or right action

Morals: relating to principles of right and wrong in behavior; expressing or teaching a conception of right behavior; conforming to a standard of right behavior

Prudent: marked by wisdom or judiciousness; shrewd in the management of practical affairs

Responsibility: the quality or state of being responsible; as moral, legal, or mental accountability

Values: something (as a principle or quality) intrinsically valuable or desirable

Virtues: conformity to a standard of right; a particular moral excellence
(Merriam-Webster Online Dictionary)

It is important that teachers offer opportunities to explore, argue about, and practice with this vocabulary, connecting abstract ethical discussions to concrete practices. However, before teachers share the definitions above with students, they should engage students in defining the terms based on their own understandings (see Appendix A). The inherent slipperiness of language is immediately evident. Emphasizing the importance of what students bring to the discussion creates a space in which open, honest, and contextual analysis of ethical issues can occur. The goal of establishing a common or accepted vocabulary creates a strong foundation for later use.

Learning an Ethical Method

After students build, refine, and assimilate a vocabulary for considering ethical issues, they need a method that they can apply to ethical decision making. Students know that whenever groups undertake the discussion of ethical decision making, personal interests and values are bound to come into conflict. Learning to effectively communicate and negotiate those differences is an essential part of becoming a professional.

It is often helpful in such discussions to let students provide examples from their own experiences: Where have they been a witness to, or experienced, ethical dilemmas that have been caused or exacerbated by communicative practices? When have they seen ethical conflicts resolved by effective communication?

In these discussions, teachers should note that classroom communications that end with disagreement are not automatically ineffective. Individuals or groups can disagree and at the same time communicate effectively. As employees, students will be asked to express their opinions about ethical issues as well as write documents that identify the best position for a company or organization to take on a given issue. Analyzing and determining the central elements of an ethical case is an important skill that has both personal and professional ramifications.

While assisting students in ethical situations, it is helpful to offer them a method of analysis that can serve them in both personal and professional situations. For the assignments provided in this chapter, I find Lori Allen and Dan Voss' (1997) method of **"value analysis"** (pp. 15-46) is most helpful. There is a significant overlap between the vocabulary terms provided above and the key terms used in Allen and Voss' methodology.

Allen and Voss (1997) define value analysis as "the ability to distinguish right from wrong" (p. 16). A value is "something we believe in" (p. 17), but not all values are ethical (Dombrowski, 2000, p. 7). **Allen and Voss define ethics as "doing what is *right* to achieve what is *good*"** (p. 17). The real strength of value analysis is its assumption that values can be both ethical and nonethical, and its recognition that terms like "good" and "right" can be interpreted in complex ways. Understanding the complicated nature of ethical deliberations provides a good point of departure for focusing classroom discussion.

Allen and Voss contend that **"value analysis is an extremely effective approach to resolving ethical conflicts because it is sufficiently structured to impose** order upon an often confusing morass of issues, yet sufficiently flexible to acknowledge the reality of *gray* when it comes to such issues" (p. 20) (emphasis in original). Discussing the specific points and terminology of the value-analysis method helps students further integrate the vocabulary and practice of ethical decision making (see Appendix B).

Negotiating Value Conflicts

Perhaps the most profound learning in technical-communication class will occur when students encounter disagreements about values that make ethical decision making difficult. Allen and Voss (1997) anticipate this difficulty in their ethical method, offering two models for thinking about values: the concentric-ring model and the hierarchal model. It is helpful to remind students that "there is no one formula for emplacing values within . . . any model" (p. 37).

The **concentric-ring model** recognizes three important sources of personal values:

- **Core Values** might include the Golden Rule, honesty, love of family (p. 24).
- **Authority Values** might include political affiliation, national organization membership, and corporate loyalty (p. 24).
- **Peripheral Values** might include personal preferences about clothing or fashion, or free time usage (p. 24).

Allen and Voss argue that when there is a disagreement in values, core values will often be a deciding factor. Teachers will want to point out that emotional responses to ethical conflict can further complicate discussions of decision making, both in the classroom and in the workplace.

The second model, **the hierarchal model**, asks student to identify the values that characterize an ethical conflict and sort them within the following hierarchical listing, thus identifying those issues that should be of the highest concern in resolving a dilemma (Critical Shared Values) and those that should be of the least concern (Personal Preference Values):

- **Critical Shared Values** include honesty, respect for property, keeping promises, and respect for the law (p. 27).
- **Important Shared Values** include family, hard work, loyalty, and compassion (p. 27).
- **Conscientiously Shared Values** include religion, political parties and "the bottom line" (p. 27).
- **Personal Preference Values** include sports, music, and diet (p. 27).

Even a brief consideration of these two models illustrates some striking similarities—both recognize the complexity of making decisions about ethical issues on the basis of values. Students will want to try out each model to see how it might help them in making ethical decisions.

APPLYING ETHICAL METHODS

Applying a method to an ethical dilemma offers students a conceptual and procedural structure for consistent decision making on ethical issues. This consistent procedure helps students establish a certain sense of decorum in their discussions of ethical issues. Objectivity, undoubtedly, is difficult to establish in connection with ethical conflicts, but using a consistent set of processes can help show students a way of tackling ethical dilemmas in a variety of situations.

A common vocabulary helps keep classroom exchanges focused on the issues at hand. With this vocabulary, students often find that they can discuss the ethical issues they have encountered in the workplace, at home, or at school, and bring these terms to bear on conflicts. By using Allen and Voss' method of analysis and employing the vocabulary it involves, students will also become increasingly effective in understanding the import of the terms that are used to discuss ethical conflicts and recognize how these terms can affect audiences and stakeholders.

Finally, this chapter offers teachers a case study to provide students practice with both vocabulary and the value-analysis method. Analyzing the case using Allen and Voss' method offers students an opportunity to identify some of the ethical choices and problems in a particular situation. The goal of the assignment is to provide students with additional practice in determining the values, interests, and needs of the various parties involved. This kind of work not only helps students become increasingly aware of how different people might respond to any communicative efforts within such situations, but helps students understand the complexities that are common when applying ethical methods.

CONCLUSION

The goal of this chapter is to supply teachers and students with a basic vocabulary for use in exploring ethical conflicts and making ethical decisions, two conceptual frameworks for understanding the complex nature of values and their relationship to decision making and a consistent method of analysis to apply to ethical conflicts. Most students who take the introductory technical-communication course will go on to work on a variety of group projects in other courses. Providing students with a vocabulary for discussing ethical conflicts, useful frameworks for understanding the role of personal values, and an effective method for resolving conflict in an ethical manner is invaluable. Considering the current state of ethical behavior in business and industry sectors, these skills and under-standings couldn't be more timely.

APPENDIX A
Speaking of Ethics; Developing a Vocabulary

Purpose

The purpose of this assignment is to provide you with the basic vocabulary for speaking about and discussing ethical issues. You might be tempted to use a dictionary or another source to define the following terms, but include you own thoughts about each term.

What You Must Do

Please write out a well-constructed and complete definition of each term. Think about your life experiences and how those experiences might inform what you know, feel, or believe about each term.

Behavior:

Code of Conduct:

Ethics:

Ethos:

Fairness:

Justice:

Morals:

Prudence:

Responsibility:

Values:

Virtues:

How You Will Be Evaluated

Please bring this list to class and be ready to discuss the definitions. Your participation in the class discussion is important. Your contribution to the discussion will help us form a working definition for each term. Therefore, you will be evaluated on your contributions to the classroom discussion and how you demonstrate your own critical thinking about these terms.

Understanding how ethical decisions are made comes through personal interaction and discussion of the issues at hand. Understanding and being able to employ appropriate language is the first step in becoming more knowledgeable about ethics. It is important to respect the beliefs of others, even when you find yourself disagreeing with them. Effective communication is both respectful and ethical. Thank you for your contributions to this important discussion.

APPENDIX B
Applying a Method of Ethical Decision Making

Purpose

Ethics is a complex field with multiple methods or strategies. The purpose of this lesson is to offer one particular method you can use when analyzing ethical dilemmas. Known as Value Analysis (*Ethics in Technical Communication: Shades of Gray*) (1997), one strength is its consistent terminology. A second strength is its straightforward organization: it has six well-defined steps. Finally, while the method is simple to navigate and understand, it is rigorous enough to be applied to complex ethical dilemmas.

What You Must Do

Identify a situation in which you or someone you know had to choose between several different ways of acting, things to do, or solutions. This situation could have happened in the classroom, the workplace, or at home. It is not necessary to be involved, but be sure to choose one with which you are familiar. Choose a decision and a situation about which several different people had very different takes on what to do. Ethical dilemmas usually occur when there are conflicting issues or values.

Read the information about the Value Analysis method in the sections that follow. Then, write a 2–3 page paper in which you apply the method to the dilemma that you selected, detailing your thinking at each step. This paper should be well organized and cohesive. One way to organize the paper is to use each step as a subheading.

After you have applied the method to the ethical conflict that you faced, write a final one-page reflection about the method. Consider the following questions as part of your reflection:

- After you have written about each step, consider whether there is a reoccurring theme that seems especially apparent to you. What is it?
- How does that theme or point relate to what you currently know or understand about the topic of ethics?
- Does Value Analysis seem to be a viable way to examine ethical dilemmas? Why or why not?
- What is most helpful about learning an ethical method? Can you think of any times where this method might have been helpful to you in the past? How so?
- Would you be inclined to use the concentric-ring or hierarchical method for rating values? Why?

The Method: Value Analysis

There are six steps to Allen and Voss' ethical method. While the steps are actually quoted from their book, *Ethics in Technical Communication: Shades of Gray* (1997). The elaboration of each step is a paraphrase of their explanation, with an emphasis on all kinds of communication. To assist you in analyzing such time, Allen and Voss have created a process by which you can determine how you will rank those values.

Step 1: Define the issue and identify the stakeholders.

Defining the issues means we must understand the audience with whom we are dealing. This is typical because effective writing is knowing or analyzing the audience (or as Allen and Voss refer to them: stakeholders). This is an important growth point because it necessitates carefully examining the importance of others. It requires looking beyond ourselves.

Step 2: Determine the stakeholders' interests.

After the audience (or stakeholders) is (are) identified, it is essential to determine the interest each stakeholder has in a particular issue. Allen and Voss (1997) note that it is important to look for—and consider the possibility of—hidden agendas at this point (p. 21). Therefore, the need for negotiation is already present. The recognition that we must negotiate brings a realization that we do not function in a vacuum; we are a corporal society.

Step 3: Identify the relevant values that bear on the issue.

This is the point where values come into the picture. Determining what values inform the issue for you personally and the audience is central to resolving a conflict. Values can be time-honored traditions; a value can be individual in nature and sometimes values are societal in nature. A value is often our understanding or belief of what constitutes whether something is good and bad. Values often determine our course of action.

Step 4: Determine the value and interests that are in conflict.

Determining values that are in conflict can be simple or complex. How do those values affect the stakeholders' interests? An orderly listing of values and interests begins to organize the issue, but more importantly we are compelled to see the connection between values and interests. An interest is something that helps give our life purpose and order in our everyday actions. A good way to demonstrate this is to say our interests help us decide what we will do.

Step 5: Apply a model to rank values according to importance, to weigh the values and interests that are in conflict.

While there are a variety of models that can be used in any situation, the model used here is based on relationships, called a "Concentric-Ring" model or on a vertical model, called a "Hierarchal" model (Allen & Voss, 1997, p. 23). Asking whether values outweigh interests is an important question. Values are a central part of both our human and corporate identity, but they are also the foundation of an ethical society.

Step 6: Resolve the conflict in favor of the higher (more important) value.

The final step is to resolve the conflict; this is accomplished by determining the higher value at stake. There are times that there is no clear-cut winner when it comes to the values at issue. This is why using one of the models discussed next is a necessary step. Developing this consciousness is central to technical-communication professionals, being both practical and ethical in their profession.

When there are conflicting values, ethical dilemmas are much more complicated. Allen and Voss offer two methods for considering and resolving conflicting values: the Concentric-Ring or Hierarchal model. You should choose which model will best serve you in thinking about the conflict you have chosen.

The Concentric-Ring Model

Allen and Voss (1997) note that there is no one formula for emplacing values within . . . any model (pp. 23-25). Human beings simply don't always agree on values. To address this complexity, Allen and Voss create three levels or rings of values:

- Core Values might include the Golden Rule, honesty, love of family, as well as religious beliefs.
- Authority Values might include political affiliation, national organization membership, and corporate loyalty.
- Peripheral Values might include personal preferences about clothing or fashion, or free time usage and recreational pursuits (p. 24).

When there is a disagreement in values, the core values are often the deciding factor. The real complexity appears when core values of different stakeholders conflict. This is a point where emotion often becomes part of the discussion. Of course emotional response can further complicate the process. It is not difficult to think of scenarios where this occurs. You will be asked to consider a scenario at a later time.

The Hierarchal Model

The second model, the Hierarchal model, encourages you to identify the values that characterize an ethical conflict and sort them within the following hierarchical listing, thus identifying those issues that should be of the highest concern in resolving the dilemma (Critical Shared Values) and those that should be of the least concern (Personal Preference Values):

- Critical Shared Values include honesty, respect for property, keeping promises, and respect for the law.
- Important Shared Values include family, hard work, loyalty, and compassion.

- Conscientiously Shared Values include religion, political parties, and "the bottom line."
- Personal Preference Values include sports, music, and diet (Allen & Voss, 1997, p. 27).

Even a quick consideration of both models illustrates some striking similarities, therefore deciding which model to employ is most likely a personal preference. Decisions will depend on how clearly a person believes the model will help them in their analysis. In either case, each model addresses the intricacy that ethical conflicts often have. This is that gray area. Again, Allen and Voss offer an important point: confronting the "grayness" of a situation is imperative. The point of employing one of the two models to value analysis is to make a difficult but necessary choice.

How You Will Be Evaluated

I will use the following criteria to evaluate your performance on this assignment:

- How well you have thought and reflected on each of the Value Analysis steps
- How well you have organized the paper so that your careful analysis of each step and its significance to method is apparent
- How well you have demonstrated that you have begun to assimilate ethical language in your writing
- How well you have answered the questions posed for the reflection section
- How free your paper is of grammar and spelling errors

APPENDIX C
A Case Study

Purpose

Case studies allow readers to examine real life scenarios and determine what they think about the course of action taken in the given situation.

Read the following case study and apply the Value Analysis method to the ethical conflict it describes:

- Identify the issues and the stakeholders.
- Determine the stakeholders' interests.
- Identify the relevant values that bear on the issue.
- Identify the values and interest that are in conflict.
- Apply a model to rank values according to importance, to weigh the values and interests that are in conflict.
- Resolve the issue in favor of the higher value.

What You Must Do

In a 2–3 page paper, write a step-by-step analysis of the case using the Value-Analysis method. The paper should be organized and cohesive; it should use the terminology we have covered in class, and it should demonstrate that you have understood both the issues present in the case and how the method provides a response to those issues.

This paper can be done in teams of two. Working with another person allows you to consider another person's perspective from the outset.

The Case

You are the main writer for a relatively new multimedia company. The company is only three-years-old, but continued growth has more than doubled the workforce. You have been with the company from the first days when there were only four employees. Your two best college friends own the company, but they depend on you for your writing skills. Often you are asked to create necessary or even new documentation as needs arise. Some of your college classes provide some degree of expertise, but often you are asked to go beyond your comfort level or what you believe is your actual level of expertise. Your work has been helpful and generally effective.

With 10 employees and a growing client base, resources are stretched thin and co-workers are often taking work home to stay on schedule. The company has chosen to salary the workers, so taking work home does not offer extra pay, it merely keeps clients happy because their work is completed on time. Over the past six months, two separate incidences have kept an employee off the job. In one case, a co-worker needed to take time off to stay with his wife and their new child. Because of a difficult delivery and subsequent complications, it was necessary for the husband to be home for an extended period (approximately three weeks). That absence backed up projects and created extra work for the rest of the staff. While the employee was paid for the time off, when he returned he felt a sense of resentment from his co-workers.

In a second case, another worker needed to care for an aged parent who was suffering with Alzheimer's. What was initially intended to be a few days absence, ended up being substantially longer. Although the employee had sufficient vacation time to cover most of the absence, the lost time had significant consequences. Because of missed deadlines, the company lost a client. Even though both workers have returned and are back to full speed, the atmosphere at work is now strained.

This past Monday, an e-mail informed employees that management plans to create and implement an Employee Leave Policy. Management (your college friends) believes the company needs to become more structured in its relationship with the workforce. This leave policy is one of those steps. Now you have been asked to put together an initial draft of the Employee Leave Policy. You have been told to research the policies of comparable firms as well as to write a policy that adheres to both Federal and State laws. You have been asked to survey the workers to determine their attitude toward developing such a policy. Your initial inquiries with co-workers have revealed resistance. Research in comparable companies has revealed policies that are more structured. What you have learned is that most employees feel this policy is impinging on their freedom; that such a policy will drive a wedge between the managers and the workforce. Some workers have said that if the policy is enforced they will quit.

You have been asked to supply both a policy draft and legal documentation to back up the policy to management in two weeks. The managers also know there is some discontent among the "troops," so they have asked that you offer a rationale for this policy implementation. By doing so, you feel stuck in the middle, but you also understand their request. Knowing there are potential conflicts, you begin research and writing.

How You Will Be Evaluated

I will use the following criteria to evaluate your performance on this assignment:

- How well you have thought about and reflected on each of the six Value Analysis steps
- How well you have organized the paper so that careful analysis of each step and its significance to this case is apparent
- How well you have demonstrated that you have begun to assimilate ethical language in your writing
- How thoroughly you understand the stakeholders, values, interests, and conflicts
- How thoughtfully you have applied the Concentric-Ring or Hierarchal Model to resolve the value conflicts
- How cohesive and clear your paper is
- How free your paper is of grammar and spelling errors

REFERENCES

Allen, L., & Voss, D. (1997). *Ethics in technical communication: Shades of gray.* New York: Wiley and Sons.

Bovée, C. L., Thill, J. V., & Schatzman, B.E. (2003). *Business communication today* (7th ed.). Upper Saddle River, NJ: Prentice Hall.

Codes of Conduct. (2005) *Institute of Business Ethics.* Retrieved April 26, 2005, from http://www.ibe.org.uk/codesofconduct.html.

Company Background. *Info about Gateway* Retrieved April 24, 2005 from http://www.gateway.com/about/news_info/company_background.shtml.

Dombrowski, P. (2000). *Ethics in technical communication.* Boston, MA: Allyn and Bacon.

Dragga, S. (2001). Quest editor's column. *Technical Communication Quarterly, 10*(3), 245-249.

Dubinsky, J. (2004a). Becoming user-centered, reflective practitioners. In J. Dubinsky (Ed.), *Teaching technical communication: Critical issues for the classroom* (pp. 13-14). Boston, MA: Bedford/St. Martin's.

Dubinsky, J. (2004b). Guest editor's introduction. *Technical Communication Quarterly, 13*(3), 245-249.

Ebbers faces prison after conviction in Worldcom scandal. (2005). *The all I need news.* Retrieved April 26, 2005, from http://www.theallineed.com/news/0503/151857.htm.

Espie, S. (2002). Why wasn't Enron's code of conduct enough? *Ethics Resource Center.* Retrieved April 26, 2005, from http://www.ethics.org/nr0318enron.html.

Ferrell, O. C., & Ferrell, L. (2002). Business ethics and social responsibility: How to improve trust and confidence in business. Retrieved April 23, 2005, from http://www.e-businessethics.com/ocf_old/Teach_BusEthicsPPTs/importance%20of%20 tchg%20ethics.ppt#258,3,Corporate Responsibility Crisis.

Gateway Homepage. (2005). *Gateway.* Retrieved April 24, 2005, from http://www.gateway.com/index.shtml.

Gateway to close retail stores. Press Release. (2004). *Gateway Inc.* Retrieved April 24, 2005, from http://www.gateway.com/about/news_info/press_release.shtml.

Global execs agree to fight corruption. (2004). *Engineering News-Record, 252*(5), 15. Accessed April 26, 2005, from Ebsco Host.

Hitt, G., & Schlesinger, J. (2002). Stock options come under fire in the wake of Enron's collapse. *Dow Jones and Company.* Retrieved April 25, 2005, from http://accounting.cba.uic.edu/Enron/Options,%20401K/Stock%20Options%20Come% 20Under%20Fire%20In%20the%20Wake%20of%20Enron's%20Collapse%20-%20Mar %2026%2002%20-%20WSJ.htm.

Johnson, R. (1998). *User-centered technology: A rhetorical theory for computers and other mundane artifacts.* Albany, NY: SUNY Press.

Lauria, J. (2003). Business ethics education initiative. *K-State business: A college for the changing world.* Retrieved April 24, 2005, from http://www.cba.k-state.edu/departments/ethics/docs/thebusiness.htm.

Markel, M. H. (2001). *Ethics in technical communication: A critique and synthesis.* Westport, CT: Greenwood.

May, G. S. (2005). Business ethics takes on more importance as business scandals make headlines. *George S May International Company.* Retrieved April 28, 2005, from http://ethics.georgesmay.com/6.htm.

Miller, C. R. (1979). A humanistic rationale for technical writing. *College English, 40*(6), 610-617.

Miller, C. R. (2004). What's practical about technical writing? In J. Dubinsky (Ed.), *Teaching technical communication: Critical issues in the classroom* (pp. 154-164). New York: Bedford/St. Martin's.

O'Neill, P. (2002). Corporate scandals spotlight need for ethics training. Retrieved April 23, 2005, from http://www.natcath.com/NCR_Online/archives/080202/080202o.htm.

Pellegrini, F. (2005). For Enron, Washington may have been a bad investment. *Time Online Edition.* Retrieved April 23, 2005, from http://www.time.com/time/business/article/0,8599,193907,00.html.

Russell, D. R. (2004). The ethics of teaching ethics in professional communication: The case of engineering publicity in the 1920s. In J. Dubinsky (Ed.), *Teaching technical communication: Critical issues in the classroom* (pp. 165-186). New York: Bedford/St. Martin's.

Schultz, B. (2004). Ethics under investigation: Network executives demand proof that vendors adhere to the highest business-conduct standards. *Network Signature Series.* Accessed April 27, 2005, from Ebsco Host 12918573pdf.

Steele, B., & Black, J. (2000). Ethics codes and beyond. *American Society of Newspaper Editors.* Updated October 25, 2000. Retrieved April 26, 2005, from http://www.asne.org/kiosk/editor/99.feb/steele1.htm.

Strauss, G. (2003). CEO's fat checks belie troubled times. *USA Today Online.* Updated June 4, 2003. Retrieved April 26, 2005, from http://www.usatoday.com/money/companies/management/2002-09-30-cover-cms.htm.

Webley, S., & Le Jeune, M. (2005). Corporate use of codes of ethics. *Institute of Business Ethics.* Retrieved April 27, 2005, from http://www.ibe.org.uk/latestpub.htm.

Word Definitions. Retrieved April 17, 2005, from http://www.merriam-webster.com/.

Yeaman, A. R. J. (2002). Professional ethics for technology. *Tech trends, 48*(2), 11-15.

CHAPTER 13

Select, Interpret, Produce:
A Three-Part Model for Teaching
Information Graphics

Karla Saari Kitalong

> **OUTCOME:** Students will critically and competently select, interpret, and produce information graphics for use in technical documents.

Information graphics—visual representations of quantities, relationships, or processes—are ubiquitous in business and technical contexts as well as in everyday life. Among their crucial functions is to help us make sense of the "overwhelming onslaught of raw data" (Wurman, 2001, p. 6) that comes to us on any given day. **Although writing teachers have traditionally been most concerned and comfortable with words, technical communication teachers are expected also to teach information graphics, which require not only verbal literacy, but also, in approximately equal measure, design sensibility, software competence, and numerical fluency.** In fact, the ability to select, interpret, and produce effective information graphics can be seen as related to "numeracy," or "mathematical literacy," "the aggregate of skills, knowledge, beliefs, patterns of thinking, and related communicative and problem-solving processes individuals need to effectively interpret and handle real-world quantitative situations, problems, and tasks" (Curry, Schmitt, & Waldron, 1996).

Most current technical-communication textbooks include a chapter or two on information graphics. Software mechanics are rarely included in these chapters, but most include guidelines for selecting appropriate forms (bar chart, pie chart, scatter plot) for presenting particular kinds of data (Anderson, 2003, pp. 291-319; Lannon, 1999, pp. 261-262). Most also advise communicators to emphasize trends or relationships without introducing bias or distortion (Lannon, 1999, pp. 294-297).

Increasingly, textbooks associate appropriate graphics with a document's usability and persuasiveness (Anderson, 2003, p. 265) and suggest how cultural differences can be accounted for in selecting and presenting graphics (Anderson, 2003, p. 286). All of these design guidelines are important and worthy of discussion in the technical-communication classroom.

In this chapter, I argue however, that standard guidelines are only a start. I discuss a pedagogy of information graphics that relies upon three interrelated literacies: interpretation, selection, and production. By assigning work that elicits, crystallizes, and then questions students' existing information-graphics knowledge, I hope students understand that **information graphics must be understood not as neutral and efficient ways to present information but as bearers of narratives. Information graphics tell stories: they convey important and complex information about cultural trends, relationships, and values (McNelis, 2000).**

WHAT BUSINESS AND INDUSTRY TELL US

Information graphics are ubiquitous and hence important to discuss in technical-communication courses. A review of practitioners' literature concerning information visualization reveals five observations of interest to technical-communication teachers.

- Employees are increasingly called upon to interpret, select, and produce effective and appropriate information graphics.
- Such information graphics are believed to be easy to create with today's software.
- Textual literacy is believed to be declining; we are embroiled in what is called the "Age of the Image."
- As a culture, Americans lack both textual or verbal literacy and sophisticated mathematical literacy.
- Although images are increasingly needed for effective communication, universal rules are believed sufficient to create such images.

Employees Need Skills to Enable Them to Interpret, Select, and Produce Effective and Appropriate Information Graphics

Human Resources and Skills Development Canada (HRSDC), which aims to "ensure that Canadians have the skills and knowledge required for the workplace," rates nine essential skills for successful employees. One of these nine skills is document use, which "refers to tasks that involve a variety of information displays in which words, numbers, icons and other visual characteristics . . . are given meaning by their spatial arrangement" (Human Resources, 2005). HRSDC lists graphs, tables, signs, and labels among "documents used in the world of work."

Interpreting information graphics, then, is a skill required at even the most basic level of complexity. More proficient (faster, more thorough) interpretation skills are needed as an employee rises through the ranks, and the more responsible the position, the more visual analysis and production skills are required. At the higher levels, employees are expected to evaluate and synthesize information found in several documents, which seems to demand the creation of comparative tables, charts, and graphs (Human Resources, 2005).

Similarly, **the State of Connecticut's core employability skills list graphic interpretation, selection, and production skills under several key areas** (Connecticut, 2001).

- Language skills/reading: interpret abbreviations, symbols, and graphs
- Language skills/writing: create and interpret charts and graphs
- Communication skills: apply data analysis to job tasks
- Mathematical skills: apply statistical principles, identify trends from data, create and interpret tables and graphs

Other state and national standards undoubtedly include similar mandates for graphic literacy, as do specific job postings; **a keyword search at careerbuilder.com revealed over 600 positions that listed as a primary requirement some proficiency with "graphing"**; these ranged from administrative assistant and portfolio-management associate to software developer, design engineer, and senior medical writer (careerbuilder.com, 27 April 2005). Thus, information-graphics proficiency is a criterion for both human-resource planning bodies and specific employers.

Information Graphics are Easy to Create with Today's Software

Teachers may take comfort in believing, along with many in the business world, that it is no longer necessary to explicitly teach students about information graphics. This misplaced belief stems from the ease with which average users equipped with common software packages such as the Microsoft Office suite can create functional and technically sound tables, graphs, and charts. The perceived ease of use of these packages (McNelis, 2000, p. 28) encourages an "anyone can do it" mentality that ignores the complexities of producing rhetorically effective and con-textually meaningful information graphics.

Software tutorials treat both software and hardware as neutral tools. From this perspective, which understands an information graphic as a function of perception alone, interpretation and critical thinking are neither expected nor well understood (Lupton & Miller, 1999, p. 141). **In short, a simplistic emphasis on software procedures and the mechanics of data-selection criteria absolves students (and teachers) of the need to demonstrate critical technological literacy and critical thinking about information graphics, qualities that have long been privileged on the academic side of technical communication and have begun to emerge as desired employee skills in technical contexts,** as pointed out by Schmitt (2000) in a passage elaborated below.

Textual Literacy is Declining in the "Age of the Image"

Despite arguments to the contrary (e.g., Moravia, 1996), conventional wisdom holds that members of Western society—particularly the young—don't read as much or as willingly as they did even 30 years ago (Steen, 1990, p. 211). This situation is attributed for the most part to television and the Internet, and is often met with nostalgia for a time when literate people were more dependent upon text than on images. However, subscribing to the belief that the "human brain processes the world visually first and foremost," many businesses have adopted a more pragmatic view: **If visual communication is "the next megatrend," it is destined to "become core to how businesses communicate with each other and with their suppliers, customers, and . . . other stakeholders."** For the business world, then, images have been "liberated by technology from . . . historical confines" (Gerard, 2001, p. 14), increasing their importance as mechanisms by which businesses can express complex concepts, relationships, and processes, and thereby increase their market shares.

Even the most nostalgic literacy advocates acknowledge the value of images for communicating cross-culturally. However, this value must be contextualized. Stocker (1992), for example, seemingly overlooks cultural difference when he extols the ability of visuals "to overcome language barriers thereby needing no explanation to convey an exact message" (p. 68). Similarly, accountant McNelis (2000) states that "graphs are not particularly prone to misinterpretation" (p. 30).

As a Culture, Americans Lack Mathematical Sophistication

Americans are considered, by and large, to be mathematically unsophisticated; even many well-educated Americans arguably lack basic mathematical competency (Dench, 1997; Steen, 1990;). Even as overall mathematical competency is perceived to be declining, companies increasingly seek out "mathematically literate" employees.

Mathematical facility is increasingly viewed as a "generic" employability skill, rather than as a specialized professional skill (Dench, 1997, p. 190). Although not every job requires mathematical fluency, businesses faced with new "competitive pressures" are responding by emphasizing an increasing array of entry-level skills and transformed workplace practices (Dench, 1997, pp. 190, 193).

Moreover, Mary Jane Schmitt, a mathematics educator, maintains that although most math instruction promotes "rule-based math," today's businesses need workers who are competent at "gaining access to information, expressing ideas, acting independently, and bridging to the future"; in short, Schmitt advocates teaching adult vocational learners to employ math in service of the "accomplishment of a larger task" (Schmitt, 2000). In the technical communication classroom, information graphics can be seen as part of this overall trend toward contextualized mathematical fluency. **Students in technical-communication classes, in short, need to be able to understand numerical relationships and manipulate such relationships in a contextually meaningful way; at least in part, this includes being able to select,**

interpret, and produce appropriate information graphics for inclusion in technical documents.

Universal Design Rules Don't Hold All the Answers

Articles in publications from a diverse array of technical professions—accounting and finance, engineering, management information systems—seem to rely on universal rules for the design of information graphics. Tractinsky and Meyer (1999) call this the "rational approach to display design," which emphasizes the "cognitive efficacy of presentations for task performance and decision making" (p. 397). Stephen Few writes, "By simply identifying the message type and knowing the proper design solutions that correspond to that type, you'll be able to conquer most of your design choices" (2004, p. 35). Similarly, Fulkerson, Pitman, and Frownfelter-Lohrke (1999) offer three "basic guidelines for preparation of financial graphics": "Choose a graph type, prepare the graph, and make sure the graph is not deceptive" (p. 29).

In "Every Picture Tells a Story," a Web article published by Ideabook (2005), information graphics are defined as "bar and pie charts with an opinion"; however, characterizing information graphics as opinionated is not, in itself, an argument in favor of critical thinking. In fact, the author suggests that the purpose of information graphics is "to grab attention" through a "visual personality" that is usually "lighthearted." This viewpoint is clearly controversial. Moreover, in the Ideabook view, to create information graphics requires the following "manageable steps":

Step 1: Gather and present accurate information.
Step 2: Focus on a single point.
Step 3: Find the most relevant image.
Step 4: Choose a palette of colors.
Step 5: Use words sparingly.

Certainly these decontextualized steps are necessary, but they are not sufficient. They gloss over the rhetorical context; in the worst-case scenario, readers will believe they need only follow such rules in order to achieve transparent communication. We must therefore rely upon our colleagues in academic settings for a rhetorical understanding of image graphics and their use within technical-communication contexts.

WHAT ACADEMIC RESEARCH TELLS US

Given the prevalence of information graphics in even the most mundane documents, it is perhaps not surprising that a good bit of research from various quarters suggests the importance of visual communication in general and the effective selection, interpretation, and production of information graphics in particular (Dragga & Voss, 2001; Harris, 1998, 1999; Harrison, 2003;

Kostelnick, 1998; Kramer & Bernhardt, 1996). Three observations from academic research are of particular interest to technical communication teachers.

- Visuals and texts work together
- Information graphics facilitate the management of complexity
- Rhetoric is a tool for avoiding graphical deception and distortion

Visuals and Text Work Together

Robert Horn, author of *Visual Language,* asserts that "visual and verbal representation systems are coming together" (Horn, 1998, p. 5) as we move into what is increasingly termed a "visual culture" (p. 4). As Laura Gurak (1998) puts it, **"Visuals are being seen as an intrinsic component of all technical communication"** (p. 492). Tufte (1983) suggests that an emphasis on aesthetics is key, but his work also encompasses convention and perception. In fact, convention and perception are, arguably, the emphases adopted in most technical-communication textbooks.

In an often-referenced passage from *The Dynamics of Document Design,* Karen Schriver (1997, pp. 412-430) lays out five perception-oriented ways of integrating text and graphics.

> Redundant: Substantially identical content appears visually and verbally; each mode tells the same story by emphasizing key ideas through repetition.

> Complementary: The visual and verbal content are different, but both modes are needed in order to understand the key ideas.

> Supplementary: Words and pictures supply different content. One mode is dominant, providing the main ideas, while the other reinforces, elaborates, explains, interprets, or provides an example.

> Juxtapositional: Words and pictures supply different content. The meaning is created or conveyed by a tension between the modes. The meaning or implication cannot be inferred without both modes being present simultaneously.

> Stage-Setting: Words and pictures supply different content. One mode (often the visual) forecasts the content, theme, or ideas presented in the other mode.

Despite our comfort level with Schriver's solid and logical descriptions of image/text integration, focusing on perception and convention at the expense of aesthetics and interpretation limits the technical-communication work that can be accomplished with information graphics.

Information Graphics Facilitate the Management of Complexity

Agreeing with the business and industry representatives cited above, a number of academic authors maintain that **information graphics allow us to manage and**

interpret vast amounts of interrelated and interdependent data in order to make both everyday and complex workplace decisions.

Self-styled information architect Richard Saul Wurman (2001), in his book, *Information Anxiety 2*, advocates that we strive to "understand the difference between data and information," so as to "reject the vast majority of the overwhelming onslaught of raw data," and ultimately "give ourselves permission to seek out and accept only that information which applies to our interests" (p. xxvii). That is, he encourages us to tell ourselves stories about information that make it relevant to us. **Wildbur and Burke (1999) concur, suggesting that information graphics be designed to help users "extract only that information which they need for a given purpose" (p. 6).**

Barbara Mirel's (2004) work on complexity represents an ambitious effort to understand information graphics; in extensive workplace-contextual inquiries, she studied, among other things, "how users move from indeterminacy to resolution" (p. xix), and her book recommends specific ways in which software designs can help them do so. Many of her recommendations directly inform the design of information graphics. **Users need to be able to "conduct complex queries visually," "keep the data context visible during complex queries" (p. 343), "put data into the right graphic display" (p. 349), and access graphic views "that easily adapt to users' arranging and rearranging needs" (p. 350).** Her contextually grounded work foretells how information graphics may be incorporated in the workplace of the future.

Mind the Rhetorical Context to Avoid Deception and Distortion

Although business and industry publications allude to the potential for deception and distortion in information graphics, they typically tie such ethical concerns to the technical accuracy of the graphic. Our business colleagues appear to ignore audience, purpose, and context, the rhetorical situation. For example, the Ideabook (2005) article apparently does not take rhetorical situations into account; if it did, it would not suggest that graphics always afford a "lighthearted" view of information. Depending upon the specific rhetorical situation and the information needs of the readers, "lighthearted" graphics could be totally inappropriate. Moreover, **many of academia's most respected theorists, including Edward Tufte, disdain the gratuitous use of images in information graphics.** John Bryan (1995) has developed a taxonomy that exhaustively illustrates all the different ways to distort data in displays. He writes, "All seven types of distortion ultimately result from the complex interplay of reader expectations, graphing conventions, and visual perception" (p. 130). The seven types of distortion are all described as "manipulations": Manipulation of scale ratios, of the second dimension, of the third dimension, of color, of composition, of symbolism, and of affect. The latter, **manipulation of affect, occurs "when the graph actively invites the reader to interpret the data through a filter of emotion"** (p. 176), often by introducing seemingly gratuitous or decorative images that may alter the tone and message of the graph.

"Chartjunk" is Tufte's name for images that serve only as decoration. He defines chartjunk as "non-data-ink or redundant data-ink"; that is, "ink that does not tell the viewer anything new" (Tufte, 1983, p. 107). He acknowledges that "principles of artistic integrity and creativity are invoked to defend—even to advance—the cause of chartjunk." Tufte concludes, however, by asserting, "There are better ways to portray spirits and essences than to get them all tangled up with statistical graphics" (p. 107). In short, Tufte might agree that designing information graphics requires following the steps outlined in the Ideabook article, but he would call for a much deeper contextualization of the steps within a framework that is, at base, rhetorical, although he does not describe it that way.

Tufte's denigration of "chartjunk" can be viewed as good news for technical communicators, given that artistic competence is not at present a core job requirement. If Tufte is correct, then we adhere to his mandate when we define effective information graphics as unadorned tables, charts, and graphs. Indeed, a number of academics invoke Tufte's intolerance of gratuitous graphics in their arguments, reinforcing the dichotomy between the how-to discourse of the practitioner and the why discourse of the theorist (Harris, 1998, 1999; Kramer & Bernhardt, 1996; Kostelnick, 1998; Mirel, 1998).

In "Cruel Pies" and "Hiding Humanity," **Sam Dragga and Dan Voss (2001, 2003) maintain that technical graphics dehumanize communication when they omit verbal and especially visual details that add texture and personality.** Obviously, they do not advocate that technical communicators trivialize serious graphics with gratuitous visual details. However, they do suggest that affect is not necessarily a bad thing for information graphics, especially depictions of disastrous occurrences. "Technical communication," they write, "does not operate in an emotional vacuum." Indeed, the deliberate omission of the human element in the interest of scientific objectivity actually communicates an "incomplete picture" and potentially introduces a "new bias" (2003, p. 78). Nonetheless, several respondents to their work dismissed the arguments, calling them "harrow[ing to] the reader's emotions" and "cater[ing] to visual illiteracy" (*Half Baked Pies,* 2002). In their response to the comments, Dragga and Voss write that although one critic

> reveals a reliance on the *verbal* component of communication to carry the weight of a humanistic perspective . . . we believe the *visual* component of communication also has the capacity to convey a sensitivity to human subjects, and that . . . technical communicators . . . have a moral responsibility to exercise that capacity by finding appropriate graphic solutions (*Half Baked Pies,* 2002).

Thus, academic technical-communication researchers, operating from a rhetorical perspective, have added a number of complex layers to the question of how best to incorporate information graphics into technical documents. The remainder of this chapter further defines and situates information graphics with respect to the technical-communication service course and outlines several assignments that can be used to reinforce the key competencies, selection, interpretation, and production.

A TAXONOMY OF INFORMATION GRAPHICS

I previously defined information graphics as visual representations of quantities, relationships, or processes. **Tables, graphs, and charts are the most commonly used information graphics in technical communication, and they further function as a means of presenting "raw data, a set of actions, or a process" visually, in the form of a "model capable of revealing its essence in terms which a particular audience can grasp easily"** (Wildbur & Burke, 1999, pp. 6-7). Wildbur and Burke's definition suggests the following taxonomy of information graphics:

1. Raw data that has been turned into "information presented as an organized arrangement of facts or data" (p. 6). Examples: tables, signage systems, and maps.
2. Information that has been "presented as a means of understanding a situation" (p. 6). Examples: line and bar graphs, organizational charts, and guide books.
3. Information that has been "presented as a means of understanding . . . a process" (p. 6). Examples: process diagrams and instructional illustrations.

Wildbur and Burke's taxonomy illustrates that each category of information graphics tells a particular story about a set of data. These authors reinforce the persuasive nature of information graphics and, in so doing, remind us of the importance of paying attention to the rhetorical situation, with particular attention to audience members' needs and motivations.[1]

VISUAL LITERACY FOR
TECHNICAL-COMMUNICATION STUDENTS

Effective use of information graphics requires technical professionals to master several different kinds of visual literacies. First, technical professionals are expected to choose ready-made graphics from existing sources; this literacy, which I label selection, involves more than aesthetics and a grasp of the concepts the graphic must convey; it also includes a sense of how, when, and why to integrate text and graphics and assumes a high level of audience awareness (Schriver, 1997).

In addition, the visually literate technical professional should understand and be able to interpret a wide variety of information graphics. In fact, interpretation is a baseline literacy that is increasingly expected of even the average citizen, as evidenced by the number and complexity of information graphics that appear in daily newspapers (Lupton & Miller, 1999, pp. 152-155) and other mundane documents.

---

[1] This chapter addresses only the first two types of information graphics in Wildbur and Burke's (1999) taxonomy. The third is too complex for beginning technical communicators and requires artistic ability to accomplish.

Beyond selecting and interpreting visual representations of data, **a third literacy, which I label production, is also expected of technical professionals** (Gurak, 1998; Harris, 1998, 1999; Horn, 1998; Lupton & Miller, 1999). The complexities inherent in producing information graphics are, as I noted earlier, glossed over because with modern software it is so easy to render technically correct charts, graphs, and tables; edit or reformat them; and repurpose them for inclusion within a variety of documents.

The literature reviewed thus far leaves us with very basic questions: What should technical graphics look like? Be like? Creating information graphics can be viewed simplistically, as a step-by-step process accomplished with basic software tools (Ideabook, 2005). However, a more academic perspective suggests that **the ability to interpret, select, and produce graphics requires much more than a knowledge of software mechanics.** Kostelnick (1998) encapsulates the problem as follows: "The need to visualize vast quantities of data, the absence of a universal standard, and the speed and power that computers now afford us have resulted in several competing ideas about how to design data effectively" (p. 473). Depending upon which competing school of thought a teacher is exposed to, she may be emphasizing convention, perception, interpretation, aesthetics (Kostelnick, 1998, p. 473), or some combination of these. Which emphasis is correct? Which is most effective to teach in technical-communication-service courses? The answers to these questions depend, of course, on the rhetorical situation; as Kostelnick so aptly points out, "conflicts among the standards [for data displays] can be reconciled by allowing the rhetorical situation to guide the design process" (1998, p. 473).

A PEDAGOGY OF INFORMATION GRAPHICS

In most technical communication contexts, words continue to be privileged over images, as evidenced by the limited and predictable treatment of visuals in popular technical-communication textbooks[2] and the visually conservative layouts of most such textbooks.

For many teachers, substantive coverage of information graphics seems out of reach in a 15-week semester; part of the reason may be teachers' lack of confidence in their own numeracy and visual communication fluency. Time is undoubtedly another factor; a cursory review of technical-communication syllabi found via a Google search indicates that at best, teachers spend a few days on graphics in the typical technical-communication course. This time is often spent reviewing available graphic forms and enumerating features that constitute effective information graphics.

I can understand then why teachers, armed with textbooks that privilege words over images and laboring under institutional pressure to cover all of technical communication in 15 weeks, would simply go over the principles covered in their textbook's graphics chapter and direct students toward the design wizards embedded

---

[2] See, for example, Anderson, 2003, chapter 11; Lannon, 1999, chapter 14; Markel, 2004, chapter 14; and Woolever, 2002, chapter 4.

within Excel, Word, and PowerPoint. **The challenge for teaching information graphics lies not in teaching students to use the software but in creating assignments that encourage critical technological literacy and that guide technical-communication students' understanding of how technical visuals make meaning—how they convey messages independent of the data they display.**

Teachers' theoretical and practical limitations notwithstanding, success in technical and professional communication requires the ability to select, interpret, and produce effective visuals and to execute effective text/visual integrations that are both purposeful and rhetorically sound. The assignments that I have developed and that I outline in the remainder of this chapter can be used in a variety of technical-communication courses to teach students visual selection, interpretation, and production strategies.

ASSIGNMENT SEQUENCE

I approach the section of the course that addresses information graphics armed with the premise that every information graphic has a story to tell. Interpreting existing graphics demands that we ask three questions:

Who is telling the story?
To whom is the story being told?
To what end? To whose benefit?

The three assignments I include here, all of which can be scaled for use as in-class assignments or homework and can be accomplished individually or in groups, are primarily designed to guide students toward competency in interpreting and designing or producing information graphics. Although the assignments are not directly focused on selecting graphics, once students develop interpretation and production skills, selecting preexisting information graphics should be straightforward. Selection is a skill perhaps best assessed in students' final course projects and may be especially important in courses that focus on service-learning or client projects.

I almost never use all three assignments in the same class. The assignments should work with any textbook you have chosen. They are presented in order from least to most complex.

Interpreting Everyday Information Graphics

I ask students to prepare for the first day of the information-graphics unit by reading the relevant chapters in the textbook and by bringing to class an information graphic from their everyday interactions. They are not to actively search for this graphic; it should be a graphic that they run across as they go about their daily business.

The overall goals of this assignment are (1) to introduce students to the idea that every information graphic tells a story, and (2) to demonstrate that even the most mundane information graphics—those everyday visuals that we hardly notice—are much more complex than they appear at first glance.

It also sensitizes them to the volumes of information graphics that cross their paths on a given day.

In class, students first classify their graphic using the taxonomy provided in their textbook and answer a few questions about the graphic to reinforce what they have read. In a short presentation, I analyze my own mundane graphic, a bar chart that appears on my monthly electricity bill, and introduce several categories of analysis. They subsequently use these categories to conduct, in small groups, a deeper analysis on the graphics they provided. Assessment usually takes place as they report the results of their group work; alternative assessments could involve collecting and checking their work, assigning a follow-up memo, or providing a form for each student to fill out.

This assignment is detailed in Appendix A.

Reshaping Data for Critical Interpretation and Production: What Stories Do Data Tell?

Although it is more time-consuming than the Everyday Graphics assignment, **this assignment also clearly illustrates that arrangement and presentation of data tell a story.**

To launch the activity, I share a data table, such as the one reproduced in Figure 1. This table presents an alphabetical list of the fifty states and the District

Median income, poverty rate in your state

Nationwide, median household income remained flat last year. The poverty level for a family of four was $18,810 or less. State-by-state median household income and percentage of people living in poverty:

State	Median income	Rank	Poverty rate	Rank	State	Median income	Rank	Poverty rate	Rank
Ala.	$35,158	44	17.1%	7	Mont.	$35,399	43	14.2%	12
Alaska	$52,499	6	9.7%	41	Neb.	$41,406	27	10.8%	33
Ariz.	$40,762	30	15.4%	11	Nev.	$45,395	18	11.5%	25
Ark.	$34,248	48	16.0%	10	N.H.	$53,910	4	7.7%	51
Calif.	$50,220	11	13.4%	19	N.J.	$58,588	1	8.4%	47
Colo.	$50,538	10	9.8%	40	N.M.	$34,805	46	18.6%	4
Conn.	$56,803	3	8.1%	49	N.Y.	$46,196	17	13.5%	18
Del.	$50,583	9	8.7%	46	N.C.	$38,234	41	14.0%	14
D.C.	$42,118	24	19.9%	2	N.D.	$37,554	42	11.7%	23
Fla.	$39,871	35	13.1%	21	Ohio	$41,350	28	12.1%	22
Ga.	$42,742	23	13.4%	19	Okla.	$35,129	45	16.1%	9
Hawaii	$50,787	8	10.9%	31	Ore.	$40,319	34	13.9%	15
Idaho	$39,492	37	13.8%	16	Pa.	$41,478	26	10.9%	31
Ill.	$47,977	14	11.3%	27	R.I.	$48,854	13	11.3%	27
Ind.	$42,067	25	10.6%	35	S.C.	$38,467	38	14.1%	13
Iowa	$40,526	33	10.1%	39	S.D.	$38,415	39	11.1%	29
Kan.	$41,075	29	10.8%	33	Tenn.	$38,247	40	13.8%	16
Ky.	$34,368	47	17.4%	6	Texas	$40,674	32	16.3%	8
La.	$34,141	49	20.3%	1	Utah	$46,873	15	10.6%	35
Maine	$39,838	36	10.5%	37	Vt.	$43,697	21	9.7%	41
Md.	$57,218	2	8.2%	48	Va.	$50,805	7	9.0%	45
Mass.	$53,610	5	9.4%	44	Wash.	$46,868	18	11.0%	30
Mich.	$44,407	19	11.4%	26	W.Va.	$31,008	51	18.5%	5
Minn.	$50,100	12	7.8%	50	Wis.	$44,084	20	10.5%	37
Miss.	$32,466	50	19.9%	2	Wyo.	$43,332	22	9.7%	41
Mo.	$40,725	31	11.7%	23					

Source: Census Bureau

USA Today 8/27/04 - p.4A

Figure 1. Poverty Data Table. Published on August 27, 2004, p. 4A. From *USA Today,* a division of Gannett Co., Inc. Reprinted with permission.

of Columbia, with poverty rates, median incomes, and the state's rank in each of those classifications.[3]

When students first examine the table, they maintain that the information presented is "neutral," by which they mean that it is completely straightforward—not interpreted or "doctored" in any way. Then, I ask them how the table can be used. As expected, they observe that it is arranged so people can easily locate their own state's data, but that state-by-state comparisons are difficult. For example, Florida is 35th in median income and 21st in poverty rate; but which states are immediately below and above Florida? The answer to this question cannot easily be discerned.

As they look more closely at the table, I ask about other possible ways of arranging the information to serve different needs. What story would the data tell if the states were arranged by poverty rank? By median income? Why would a newspaper like *USA Today*, a self-described "good news paper" with the "persistent theme" of "national unity" (Lupton & Miller, 1999, p. 144) want to make state-by-state comparisons difficult for readers? In other words, what is masked in this arrangement? It doesn't take students long to begin to question the data display, to understand that no arrangement of data can truly be neutral, and to recognize that "graphics that are apparently 'objective' or straightforward may have complex inflections" (Lupton & Miller, 1999, p. 149). In fact, "unlike narrative forms, [information graphics] may be read analytically from different vantage points, yielding different insights" (Lupton & Miller, 1999, p. 152).

The assignment is detailed in Appendix B.

Visualizing Relationships and Trends in Data

The next step is for the students to produce information graphics that reveal different insights, that tell different stories. This assignment is intended to enhance students' skills and competencies as follows:

- Interpretation: Students practice reading and understanding statistical data.
- Production: Students manipulate data by altering display formats and by combining data from different sources.
- Interpretation: Students understand that juxtaposing data in different ways reveals previously unnoticed trends and relationships, and may also open up the potential for distortion and deceit.

For this assignment, I use data available every August from the *Chronicle of Higher Education* Almanac edition (see Figure 2). This educational data is technical in nature but can be understood by all of the students; it affects them as students, citizens, and taxpayers; and the implications are revealed through close reading and reflecting upon the accompanying narratives.

---

[3] Government Web sites, such as the Census Bureau site (www.census.gov), are excellent sources of such tables.

FLORIDA – DEMOGRAPHICS
Population:
State: 17,019,068 (Rank: 4)
Nation: 290,809,777

	State:	Nation:
Age distribution:		
Up to 4	6.2%	6.8%
5 to 13	11.8%	12.8%
14 to 17	5.3%	5.7%
18 to 24	8.4%	9.8%
25 to 44	27.5%	29.4%
45 to 64	23.7%	23.1%
65 and older	17.1%	12.3%
Racial and ethnic distribution:		
American Indian	0.4%	1.0%
Asian	1.9%	4.0%
Black	15.8%	12.7%
Pacific Islander	0.1%	0.2%
White	80.8%	80.7%
More than one race	1.1%	1.4%
Hispanic (may be any race)	18.1%	13.4%

Figure 2. Sample data for information graphics assignment. **Source:** *Chronicle of Higher Education* Almanac 2004-2005. Online edition available http://chronicle.com/free/almanac/2004/. Reprinted with permission.

Before class, I select some states and make several copies of their published data sets. For the purposes of this exercise, any state's data will suffice, and teachers can use any selection criteria they wish.

Each student chooses, or is assigned, a state. I don't tell them that several students are assigned the same state. Their initial homework assignment is to read carefully their assigned data set and accompanying narrative, identify interesting issues or questions the data reveal, and create at least one information graphic based on the data. Figure 3 shows an example of a typical source data table.

I begin the next class by grouping the students with others who were assigned the same state. After they have analyzed their own state's data and compared graphics, they regroup with students who worked with other states' data. Working together, the new groups create an information graphic that compares data from two states and thereby reveals additional new insights.

This assignment typically generates a lot of interest among the students; while reinforcing strategies for interpreting and producing information graphics, it also educates them on educational trends and demographic relationships from around the country.

Recently, the students "hijacked" this assignment. While working with classmates who had selected the same state's data for analysis, they became curious about how

	State:	Nation:
Educational attainment of adults (highest level):		
8th grade or less	6.7%	7.5%
Some high school, no diploma	13.4%	12.1%
High-school diploma	28.7%	28.6%
Some college, no degree	21.8%	21.0%
Associate degree	7.0%	6.3%
Bachelor's degree	14.3%	15.5%
Graduate or professional degree	8.1%	8.9%

Figure 3. Educational attainment of adults in Florida compared with national rates.
Source: *Chronicle of Higher Education* Almanac, 2004-2005. Online edition
available http://chronicle.com/free/almanac/2004/. Reprinted with permission.

various states compared on key demographic measures. They couldn't wait for the small group activity, so we created an impromptu table to compare factors such as earned bachelor's degrees, numbers of new high school graduates, high school drop-out rates, per-capita personal income, and how many residents speak a language other than English at home (see Figure 4). By probing these relationships, students generated hypotheses that they explored further in their small group discussions. Several multistate groups used this line of questioning as the context for exploring the effect of certain statistical relationships on a state's educational funding priorities.

Depending upon the time available, the groups present, post, or write a memo about their analysis. The assignment is detailed in Appendix C.

CONCLUSION

In this chapter, I have illustrated the importance of incorporating information graphics into the technical-communication-service course. **Information graphics are essential communication tools; in addition, they tell stories that may or may not coincide with the stories told by the words they accompany.** Obviously, adequately conveying all the skills and sensibilities associated with selecting, interpreting, and producing competent information graphics would take far more time than teachers can reasonably spend in a technical-communication-service course. Software mechanics are a necessary component of the work, as are the discussions that typically appear in technical-communication textbooks and practitioners' articles, in which the best type of information graphic are defined for different communication goals (Anderson, 2003, pp. 291-319; Lannon, 1999, pp. 261-262). Important, as well, are guidelines for incorporating graphics to enhance usability and persuasiveness (see, for example, Anderson, 2003, p. 265), guidelines for addressing cultural differences in the interpretation of graphics (see, for example, Anderson, 2003, p. 286), and advice on how to emphasize trends or relationships without distorting (see, for example, Lannon, 1999, pp. 294-297).

Florida	23.1
Hawaii	26.6
Idaho	9.3
Michigan	8.4
New York	28.0
North Carolina	8.0
Wisconsin	7.3

Figure 4. Percentage of citizens whose home language is not English.
Source: *Chronicle of Higher Education* Almanac, 2004-2005. Online edition available http://chronicle.com/free/almanac/2004/. Reprinted with permission.

The assignments outlined in this chapter consolidate several of these crucial skills and competencies, thereby integrating software mechanics and procedural knowledge with an awareness of the stories that information graphics tell within the context of Western technoculture. Within a rhetorically based technical-communication-service course, these assignments, or versions of them, make good use of the limited time available for the study of information graphics.

APPENDIX A
Interpreting Everyday Information Graphics

Homework to Prepare for this Assignment

1. Read the chapter(s) on information graphics in the textbook.
2. Without actively searching, locate an information graphic; for example, from one of your textbooks, a magazine or newspaper, your daily mail, an online site that you happen to visit, a flyer that you receive in the mail. Bring a hard copy of this information graphic to class with you.

Individual Work During the First Five Minutes of Class

1. Use your textbook to identify the type of graphic (bar chart, pie chart, etc.).
2. Briefly describe the rhetorical situation to which the graphic responds, and explain the facts, processes, or relationships it depicts.

Brief Lecture/Presentation: The Electric Bill Bar Chart

I present a brief analysis of a sample graphic, the Electric Bill Bar Chart. The following are points I make when I deliver the presentation:

Recently my electricity provider began including a bar graph in each month's bill.

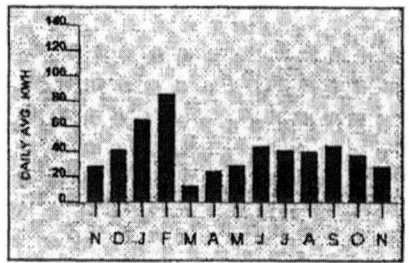

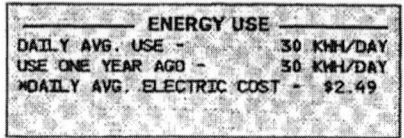

Stated Purpose: The graph tracks household power consumption for the previous year and reports the average daily cost of electricity. It clearly shows monthly and seasonal fluctuations in my power usage.

Ways to Use This Visual Information: Of course, a family could use this visual information as part of a plan to reduce overall power consumption. What other uses come to mind?

Questions Raised: Why was February's consumption higher than January's, given that February is a shorter month? Perhaps electric meters are only actually read every other month, with alternate months' bills being estimated. Thus the graphic is likely based not on actual consumption but on billed consumption.

Rhetorical Purpose of the Graphic: The graphic is largely a marketing device, emphasizing how little electricity costs and celebrating the fact that we live in Florida: power consumption in the coldest months is much higher than in the hottest summer months, so we can conclude that heat costs more than air conditioning.

Story It Tells: Here, I elicit the students' input by asking them whether $2.49 per day seems like a reasonable cost for electricity. The story the graphic tells varies depending, among other things, upon whether or not the reader is financially comfortable. The students offer other interpretations.

Analysis of Graphics Provided by Students

1. Using the categories presented in the lecture, work in small groups to analyze the information graphics you supplied.
2. Share your findings with the rest of the class when prompted to do so.

Assessment

In my course, this is typically an in-class or homework assignment. One of the following methods of assessment ensures that students document their work:

- Students give brief oral presentations summarizing their observations.
- Students complete and turn in a check sheet.
- Students post a brief response on the class web site.

Variations

The following scalable variations on this assignment are also possible:

1. For service-learning or client-based classes, students can analyze information graphics used by their partner agencies and report their findings in a memo to management, a recommendation report, or a section of the project style manual.
2. Collect student responses in the form of a quiz or in-class graded assignment.
3. Control the lessons learned by supplying information graphics that you have collected.

APPENDIX B
Reshaping Data for Critical Interpretation and Production:
What Stories Do Data Tell?

OVERVIEW

The poverty data discussed in class has been entered into an Excel spreadsheet for you (see Figure 5). This assignment involves the following steps.

1. Download the following from the class web site:
   —The Excel File
   —The *USA Today* article (electronic version)
   —The scanned data table
2. Working with one or two classmates, read the article, analyze the data, and create a new information graphic that tells a different story or reveals different insights.
3. Be sure your graphic is clear, accurate, and free from distortion. Use the criteria from your textbook as a guide.
4. You will have the remainder of today's class time to work on this project. You may continue to work on it outside of class. The final is due at the beginning of class on [the last day of our unit on graphics].

Assessment

Grading is based on

• Adherence to the graphics checklist in your textbook
• The appropriateness of the graphic format you have chosen
• Your group's presentation of your new graphic
• An in-class writing that you will do on [the last day of our unit on graphics].

In-Class Writing

On the day the groups turn in or present their information graphics, students individually respond to the following questions:

State	Median Income	Rank	Poverty Rate	Rank
Ala.	$35,158	44	17.1%	7
Alaska	$35,158	6	9.7%	41
Ariz.	$40.762	30	15.4%	11
Ark.	$34,246	48	16.0%	10
Calif.	$50,220	11	13.4%	19
Colo.	$50,538	10	9.8%	40
Conn.	$56,803	3	8.1%	49
Del.	$50,583	9	8.7%	46
D.C.	$42,118	24	19.9%	2
Fla.	$39,871	35	13.1%	21
Ga.	$42,742	23	13.4%	19
Hawaii	$50,787	8	10.9%	31
Idaho	$39,492	37	13.8%	16
Ill.	$47,977	14	11.3%	27
Ind.	$42,067	25	10.6%	35
Iowa	$40,526	33	10.1%	39
Kan.	$41,075	29	10.8%	33
Ky.	$34,368	47	17.4%	6
La.	$34,141	49	20.3%	1
Maine	$39,838	36	10.5%	37
Md.	$57,218	2	8.2%	48
Mass.	$53,610	5	9.4%	44
Mich.	$44,407	19	11.4%	26
Minn.	$50,100	12	7.8%	50
Miss.	$32,466	50	19.9%	2
Mo.	$40,725	31	11.7%	23
Mont.	$35,399	43	14.2%	12
Neb.	$41,406	27	10.8%	33
Nev.	$45,395	18	11.5%	25
N.H.	$53,910	4	7.7%	51
N.J.	$58,588	1	8.4%	47
N.M.	$34,805	46	18.6%	4
N.Y.	$46,195	17	13.5%	18
N.C.	$38,234	41	14.0%	14
N.D.	$37,554	42	11.7%	23
Ohio	$41,350	28	12.1%	22
Okla.	$35,129	45	16.1%	9
Ore.	$40,319	34	13.9%	15
Pa.	$41,478	26	10.9%	31
R.I.	$48,854	13	11.3%	27
S.C.	$38,467	38	14.1%	13
S.D.	$38,415	39	11.1%	29
Tenn.	$38,247	40	13.8%	16
Texas	$40,674	32	16.3%	8
Utah	$46,873	15	10.6%	35
Vt.	$43,697	21	9.7%	41
Va.	$50,805	7	9.0%	45
Wash.	$46,868	16	11.0%	30
W.Va.	$31,008	51	18.5%	5
Wis.	$44,084	20	10.5%	37
Wyo.	$43,332	22	9.7%	41

Figure 5. Data from *USA Today* (2004) poverty graphic entered into a spreadsheet.

- What comparisons can now be made that could not be made with the published arrangement of information? What new relationships can be seen that could not be seen before?
- What effect does rearranging the data have on the story?
- What story did your group want to tell that could not be told, given the nature of the data and the characteristics of the software?

Variations

- Have each group create a memo of transmittal to accompany their graphic, in which they address their work process and answer the questions outlined above.
- Here are some alternate instructions to use if you are teaching an online course.

> Post your information graphic to the designated discussion area. In your discussion posting,
> —Explain the story now told by the data as a result of your manipulation.
> —Discuss the design decisions you made.
> —Include the names of the people in your group.

APPENDIX C
Visualizing Relationships and Trends in Data

In class, each student receives a photocopy of one state's annual education data, as reported in the Almanac issue of the *Chronicle of Higher Education* (published each year in August) or another complex, multifaceted data set relevant to their students' specific needs.

Homework to Prepare for This Assignment

Individually, using the data you received in class, create one graph, chart, or table to illustrate some aspect of education in "your" state. You can do this with Excel or another spreadsheet program or perhaps with basic data display tools in PowerPoint or Word.

Write a short explanation to anchor your display. Bring printouts of the graphic and explanation to class with you next time.[4]

In Class

1. During class time, you will first be grouped with others who studied the same state's data. Studying each other's illustrations will help you further to understand the story of higher education in your state.

---

[4] I teach in a computer classroom, so students can post their work to their group discussion forum. This enables them to share materials before they come to class and download them for further in-class manipulation.

2. I will then ask you to form new groups with people who studied other states' data. As a group, you will create one new data display that brings together data from all the states represented in your group.
3. You may continue to work on your display outside of class; be prepared to present it briefly the next time we meet.

Assessment

Your grade on this assignment will be based on the following criteria:

- Individual:
  a. Your graphic's adherence to the graphics checklist in your textbook and the appropriateness of the graphic format you have chosen.
  b. The narrative you write to accompany your graphic, which should explain the story the graphic tells, how it differs from the story told in the original data set, and what possible distortions you have introduced with your manipulation.
  c. An individual in-class writing concerning your group experience.
  Note: This in-class writing includes a statement about the individual's contribution to the group effort, a description of data "stories" the group considered but didn't complete, and a discussion of the limitations of the software for telling certain kinds of data "stories."
- Group: Adherence to the graphics checklist in your textbook and the appropriateness of the graphic format you have chosen.

Variations

1. When I teach this class in a "hybrid" modality (some face-to-face and some online meetings), I design group, individual, and interactive work for each modality.
2. This assignment carries a greater point value than the other two assignments. It is also more difficult to integrate because it is relatively time consuming and involves specialized data. It works exceptionally well in courses devoted to writing in a specific discipline, service learning, or client-based projects.

REFERENCES

Anderson, P. (2003). *Technical communication: A reader-centered approach* (5th ed.). Belmont, CA: Thomson.

Bryan, J. (1995). Seven types of distortion: A taxonomy of manipulative techniques used in charts and graphs. *Journal of Technical Writing and Communication, 25*(2), 127-179.

Connecticut Department of Labor. (2001, November 13). Connecticut's school-to-career core academic and employability skill requirements. http://www.ctdol.state.ct.us/schtocar/coreskill.htm. Accessed April 27, 2005.

Curry, D., Schmitt, M. J., & Waldron, S. (1996). *A framework for adult numeracy standards: The mathematical skills and abilities adults need to be equipped for the future.* http://www.literacynet.org/ann/framework-full.html. Accessed November 5, 2006.

Dench, S. (1997). Changing skill needs: What makes people employable? *Industrial and Commercial Training, 29*(6), 190-193.

Dragga, S., & Voss, D. (2001). Cruel pies: The inhumanity of technical illustrations. *Technical Communication, 48*, 265-274.

Dragga, S., & Voss, D. (2003). Hiding humanity: Verbal and visual ethics in accident reports. *Technical Communication, 50*, 61-82.

El Nasser, H., & Overberg, P. (2004, August 27). Americans hang on to bigger, better lifestyles. *USA Today*. News section, p. 4a.

Few, S. (2004, September 18). Eenie, meenie, minie, moe: Selecting the right graph for your message. *Intelligent Enterprise*, pp. 35-39.

Fulkerson, C. L., Pitman, M. K., & Frownfelter-Lohrke, C. (1999). Preparing financial graphics: Principles to make your presentations more effective. *The CPA Journal, 69*, 28-33.

Gerard, A. (2001). Visual communication—The next megatrend. *Electronic Publishing, 25*(8), 14.

Gurak, L. J. (1998). Toward consistency in visual information: Standardized icons based on task. *Technical Communication, 50*, 492-496.

Half Baked Pies? (Correspondence). (2002, February). *Technical Communication, 49*(1), xx-yy.

Harris, R. L. (1998, December). The power of information graphics, Part 1 of 2. *IIE Solutions*, http://www.iienet.org/membersonly/backissues/1298wellr.pdf.

Harris, R. L. (1999, January). The power of information graphics, Part 2 of 2. *IIE Solutions*, pp. 26-27.

Harrison, C. (2003). Visual social semiotics: Understanding how still images make meaning. *Technical Communication, 50*(1), 46-60.

Horn, R. E. (1998). *Visual language: Global communication for the 21st century*. Bainbridge Island, WA: MacroVU, Inc.

Human Resources and Skills Development Canada. (2005, February 28). Essential skills. http://www.15.hrdc-drhc.gc.ca/English/general/home_e.asp. Accessed April 27, 2005.

Ideabook. (2005). *Every picture tells a story*. http://www.ideabook.com/chart.htm. Accessed November 5, 2006.

Kostelnick, C. (1995). Cultural adaptation and information design: Two contrasting views. *IEEE Transactions on Professional Communication, 38*, 182-196.

Kostelnick, C. (1998). Conflicting standards for designing data displays: Following, flouting, and reconciling them. *Technical Communication, 45*, 473-482.

Kramer, R., & Bernhardt, S. (1996). Teaching text design. *Technical Communication Quarterly, 5*(1), 35-60.

Lannon, J. M. (1999). *Technical communication* (8th ed.). Boston, MA: Addison Wesley Longman.

Lupton, E., & Miller, A. (1999). *Design writing research: Writing on graphic design*. London: Phaidon Press.

Markel, M. (2004). *Technical communication* (7th ed.). Boston, MA: Bedford/St. Martins.

McNelis, L. K. (2000). Graphs, An underused information presentation technique. *National Public Accountant, 45*, 28-30.

Mirel, B. (1998). Visualizations for data exploration and analysis: A critical review of usability research. *Technical Communication, 45*(4), 491-509.

Mirel, B. (2004). *Interaction design for complex problem solving: Developing useful and usable software*. San Francisco, CA: Morgan Kaufmann.

Moravia, A. (1996, March). The image and the word. *Unesco Courier*, p. 50.

Schriver, K. A. (1997). *Dynamics of document design.* New York: Wiley.

Schmitt, M. J. (2000). Developing adults' numerate thinking: Getting out from under the workbooks. *Focus On Basics,* 4(B). http://www.ncsall.net/?id=319. Accessed November 5, 2006.

Steen, L. A. (1990). Numeracy. *Daedalus, 119*(2), 211-231. http://www.stolaf.edu/people/steen/Papers/numeracy.html. Accessed November 5, 2006.

Stocker, G. (1992). Use symbols instead of words. *Quality Progress, 35*(11), 68+5 pages.

Tractinsky, N., & Meyer, J. (1999). Chartjunk or Goldgraph? Effects of presentation objectives and content desirability on information presentation. *MIS Quarterly, 23,* 397-420.

Tufte, E. (1983). *The visual display of quantitative information.* Cheshire, CT: Graphics Press.

Wildbur, P., & Burke, M. (1999). *Information graphics: Innovative solutions in contemporary design.* London: Thames & Hudson.

Woolever, K. R. (2002). *Writing for the technical professions* (2nd ed.). New York: Longman.

Wurman, R. S. (2001). *Information anxiety 2.* Indianapolis, IN: Que.

CHAPTER 14

Listen Up!
Oral Presentations in the
Technical Communication Classroom

Patricia Freitag Ericsson

> **OUTCOME:** Students will be able to author and deliver effective oral presentations that make appropriate and rhetorically effective use of available means, media, and modalities.
> Students will develop effective strategies for listening to and evaluating oral presentations.

WHAT BUSINESS AND INDUSTRY TELL US

We hear it all the time: **speaking and listening skills are important for success. As many listening experts point out, however, hearing is not listening.** Perhaps some numbers on the importance of speaking and listening skills can help us *hear* better and *listen up*. In 2004, Randall S. and Katharine Hansen of "Quintessential Careers" cited communication skills as the number one "critical employability skill" desired by employers. They describe communication skills as listening, verbal, and written skills and quote the typical job ad line that highlights this skills, "Exceptional listener and communicator who effectively conveys information verbally and in writing" (para. 6). According to a survey by American Express Small Business Services, verbal communication skills were at the top of the list of important skills with 86% of those surveyed considering them "very important" (2000, p. 12). Listening was rated as an "extremely important" skill by 73% of the group of business leaders surveyed by Don Carstensten, former vice president for education services at ACT. These same business leaders said that only 19% of their high school graduate employees were good listeners (Purdy, 2000, para. 2). According to Kenneth Petress, a specialist in organizational persuasive strategies, **"Good listeners**

have advantages: they get promotions, they are more frequently selected for power/prestige positions, and they are often better informed than are poor listeners" (1999, para. 13).

Studies of skills needed in different job areas also emphasize the importance of speaking and listening. In their 2002 study of the critical skills needed for Information Systems (IS) professionals, Cheryl Noll and Marilyn Wilkins found that the **ability to develop and deliver effective, informative, and persuasive presentations was one of the eight "core skills" that was needed for any IS curriculum—no matter what the concentration** (Table 6). A study of the skills most needed for banking and commercial careers listed "communication skills" and "people skills" as two of the top three abilities needed. "Analytical skills," which one might assume are essential for these jobs, were reported as only "medium" in importance in the Noll/Wilkins study (2000). These studies and numerous others echo what the 1992 Department of Labor SCANS (Secretary's Commission on Achieving Necessary Skills) found: that listening and speaking are two of the six fundamental or basic skills needed for high-skill, high-wage employment. A 2000 update of this report by Johns Hopkins Institute for Policy Studies reaffirms that these skills are vital for employment and success (section F5 & F6). Dianne Schilling (2004), a specialist in designing custom training programs for business and industry, maintains that listening is a "master skill" that is critical to success. Despite listening being critical, Schilling claims **listening is rarely taught because "educators (along with almost everyone else) assume listening is tantamount to breathing—automatic"** (para. 3).

WHAT ACADEMIC RESEARCH TELLS US

Technical-communication teachers need to *listen up* to these experts and the results of their studies. If we truly *hear* them, we need to do something about emphasizing speaking and listening in technical-communication courses. Speaking and listening need to be emphasized in our courses because the traditional general education requirement of at least one speech course has experienced a decline. Almost every college and university has felt the pressure to lower the number of credits in the General Education curriculum. As a result, speech courses, the traditional site for learning about public speaking and listening, have too frequently become options rather than requirements. In 1998, the Boyer Commission Report stated that "hardly any [students] are exposed to courses or class requirements in oral communication" (p. 22). In a 2001 update, 19% of the Boyer respondents reported that oral communication skills were taught in their school's "introductory courses" and "about 30% reported that they do not offer any courses or activities to promote development of these skills" (p. 20). A 2003 report by the National Communication Association claims that universities devote "considerable attention to writing, but much less to oral communication." **According to this report, oral communication experiences have not become a priority at universities, and students have little incentive to hone communication skills unless communication courses are required in their majors** (p. 6).

According to M. A. K. Halliday (1985), oral communication has been devalued by our "literate culture," which "tend[s] not to take the spoken language seriously." Halliday claims that writing has "taken over many of the high prestige **functions** of language in our society" and that our "highly valued **texts** are now all written ones" (bold in the original, p. 97). Despite the consecration of the written word, Halliday and others argue that listening and speaking are as important in learning as are reading and writing (pp. 96-99). In the academy, written communication has long been favored over oral discourse[1] and is regularly taught at many levels in the university with a traditional first-year writing course as an almost universal requirement.[2] This first writing course is often followed by a second course in writing (often at the junior level), a writing-across-the-curriculum program, or writing in the major courses. More often than not, support for improving oral communication is paltry compared to what is provided for written communication. **While Writing-Across-the-Curriculum programs have become commonplace in higher education, similar Speaking-Across-the-Curriculum programs are much less common.**[3]

Given the importance of speaking and listening skills for success and the paucity of courses that emphasize these skills, technical-communication courses need to include these skills so that students are better prepared to succeed. Technical-communication courses that ignore speaking and listening skills are doing a disservice to their students. But perhaps teachers need not shoulder all the blame. Teachers who ignore or give short shrift to speaking and listening are following the lead of most technical-communication texts. The first problem may stem from the fact that most books published and used in technical-communication classes are "technical-writing" texts. Despite this nomenclature, many of these texts give only a nod to speaking. In a review of a dozen technical-communication texts, I found that those texts that include a mention of speaking do so in a cursory, add-on chapter, or in a few pages in a miscellaneous chapter. Without exception, these chapters are at the end of the texts. After looking at just a few of these texts, **I realized that I could locate the speaking information (if it existed) without even looking at the table of contents—I just had to look at the last chapter of the book.** The average

---

[1] The bias in favor of written communication is seen in tenure and promotion decisions where written publications are more highly ranked than oral presentations. Some universities do not consider oral conference presentations in these decisions at all. The most frequently used and highly valued student output is the written composition. Lectures are often based on written texts, and in assessments of all kinds, the written word, rather than the spoken word, is the most highly valued (Barthel, 2001, para 6).

[2] The 2001 Boyer Report update claims that "almost all" research universities have "freshman writing courses," and that 52% offer a "two-semester sequence." In addition, 32% have upper-division writing requirements and 22% have "some other way of infusing extended writing project into the undergraduate curriculum" (p. 20).

[3] Programs in Speaking or Communication Across the Curriculum have been developed at Pittsburgh University, North Carolina State University, University of North Carolina Greenboro, Eastern Illinois University, Ohio Wesleyan University, Southeast Missouri State University, Southern Illinois University Carbondale, Columbia College, and Rice University, to name a few.

number of pages devoted to oral presentations in these texts is 10. Most of these texts run at least 500 pages, indicating how little emphasis speaking is given.

Given the lack of attention to speaking and learning in technical communication textbooks, it is not surprising that these two skills, highly valued for job success, are rarely emphasized in technical-communication classes. If textbooks gave these skills the attention they deserve, perhaps teachers would also pay more attention. We need to lobby the authors of these texts to include speaking and listening. In the meantime, responsible technical-communication teachers can incorporate more on these important skills in their classes without additional pages.

How can we incorporate crucial speaking and listening skills into our technical-communication classes effectively? As with any skill that is actually valued, teaching speaking and listening skills cannot be a one-shot add-on to the technical-communication course. Speaking and listening skills must be emphasized throughout with speaking and listening assignments and helpful critiques. Because students focus on graded assignments, these assignments must be carefully evaluated and make up a meaningful portion of the course grade. Below, I recount my approaches to incorporating speaking and learning in a technical-communication course.

THE ASSIGNMENT SEQUENCE: SPEAKING IN THE TECHNICAL-COMMUNICATION COURSE

One of the most effective ways to incorporate speaking in technical-communications courses is through "repurposing." Repurposing, a common term in software development and Web design, can be adopted to the technical-communication curriculum. In web site design, repurposing content allows for several different presentations of information—with more graphic content, alternate ideas emphasized, or a format that is designed for easier printing. Many software creators promote integrated programs that allow for repurposing. An online article, "Get the Most Out of Your Powerpoint," explains how to repurpose a Powerpoint presentation so that it can be an autorun CD, a QuickTime movie, or an e-mail presentation (Bajaj, 2003). **Thinking of repurposing as recycling, repackaging, or recrafting content makes the term easy to understand.**

Repurposing content makes sense, and good teachers do it regularly, although typically do not call the process repurposing. Technical-communication assignments often require students to write brief and full-blown proposals, reports, letters, and memos all based on one body of content. In these repurposing assignments, students learn real-world skills—skills that business and professional writers frequently employ to repackage information for different purposes and audiences. **Building speaking assignments that repurpose written material is no more difficult than assigning repackaged writing assignments, although teachers must be willing to devote class time to oral delivery, instruction in oral delivery, and time for oral presentations.**

Speaking opportunities should be included throughout the course, so that students' comfort in making oral presentation builds and so that feedback from

previous speaking assignments can be used. A one-shot speaking assignment diminishes the importance of speaking skills and does not allow students to improve their speaking skills. Ideally, students should have the opportunity to give at least three individual oral presentations and one group presentation in the course of a semester. With fewer than four speaking assignments, students will devalue the importance of these assignments. More than four may demand too much class time.

The Pitch Proposal

In the technical-communication classes I teach, the first speaking experience is typically a part of the Pitch Proposal assignment (see Appendix A). Because the course includes a team project, the pitch-proposal assignment is designed to promote a team project idea in a 5- or 6-page written proposal and an oral presentation limited to 5 minutes. The proposal goal is to persuade other class members to be a part of a team. Because the authors of the pitch proposal know the audience and have a real-world goal (getting classmates to partake in their projects), there is an objective other than getting a decent grade. Students are introduced to the format of the proposal and provided with direct instruction on how written and oral delivery differ (see Appendix B).

Subsequent speaking assignments are directly linked to the technical-writing projects students are completing, but each assignment emphasizes another element of oral presentation. In one assignment, discussions focus on different types of delivery—impromptu, memorized, reading, and extemporaneous. The pros and cons of these delivery methods are considered, and the types of delivery appropriate for different situations are discussed. Extemporaneous delivery in which the speaker has prepared and practiced, but the speech is not written out, read, or memorized, is preferred. Discussing, practicing, and requiring extemporaneous delivery allows students to learn about audience feedback signals and how to adjust messages when they receive confusing or negative feedback.

Effective Use of Voice

Another assignment focuses on effective voice use (see Appendix C). Students are introduced to voice elements like pitch, rate, loudness, and variety. In-class exercises in which students practice these elements in small groups are some of the most effective and entertaining activities of the semester. As in the other speaking assignments, use of video tapes of public speakers is effective in illustrating the presentation elements being emphasized.

Oral presentation skills are additive. The skills emphasized in the first oral presentation cannot be left behind when subsequent presentations are being prepared and delivered. Because technical-communication teachers are providing instruction in so many areas, oral presentation skills emphasis cannot be comprehensive and must be chosen carefully. Careful choice and emphasis throughout the course helps students develop speaking skills that will help them succeed.

THE ASSIGNMENT SEQUENCE: LISTENING IN THE TECHNICAL-COMMUNICATION COURSE

Because listening skills have such a premium in the job world, helping students develop critical listening skills should be an element of every class, but is particularly appropriate in technical-communication classes where all elements of communication can be emphasized. Teaching listening skills might be one of the most elusive tasks in education. Listening seems natural, but the kind of listening technical-communication classes should seek to emphasize is "critical listening," which is not something "natural." In critical listening, listeners evaluate, assess, and interpret the speaker's message. According to a top-selling speech textbook, "critical listeners analyze and evaluate what they hear" and carefully "examine a message rather than accepting it at face value" (Osborn & Osborn, 2000, p. 61). Incorporating critical-listening exercises takes teacher commitment, but not nearly as much time as incorporating speaking.

The Gist

Critical listening can be introduced by first requiring students to write "gists" or tight, brief summaries of class lectures or other presentations.[4] The gist, typically no more than 20 words, must state the main idea of the presentation and the supporting details. To compose a gist, students must listen critically and take notes. Although writing a gist does not require as much analysis and evaluation as the best critical listening would demand, this exercise starts the process to reach the level of critical listening desired.

The gists are handed in by e-mail or at the beginning of the class period following the presentation and are graded ✓,✓–, or ✓+ indicating that the gist is adequate, inadequate, or very good. If the gist receives a ✓-, ideas on how it can be improved are added.

The Advanced Gist

Once students learn to listen carefully enough to write adequate gists, the bar can be raised to include more (see Appendix D). Advanced gists, which require a summary and supporting details plus evaluation of the supporting details, can be added to the course.

Peer Evaluation

Critical-listening skills can be developed during student oral presentations (see Appendix E). Too often, students do not take other student presentations seriously, so requiring critical-listening feedback from students can help alleviate that problem.

---

[4] Thanks go to Dr. Julie Smith, formerly of Dakota State University, for introducing me to the concept of gist writing. More on gist writing can be found on the Internet. A search on "writing gists" results in multiple sources, several of which teachers might find helpful. My approach to gist writing has been developed and refined through several semesters of trial and error.

During each oral-presentation session, students (at least two) can act as peer evaluators and be assigned to listen for specific details. For example, the Pitch-Proposal oral presentation has specific goals which include (1) transforming the language of the written pitch proposal into appropriate oral-delivery language, (2) providing adequate signposts for listeners to follow the presentation easily, and (3) persuading other class members to join a team. Peer evaluators look for evidence of these goals and provide written feedback on whether or not these goals were met.

As critical-listening skills develop, student evaluation can play a small role in grading. Student evaluations can be as little as 5% of the final oral-presentation grade, but many students like having their evaluations count. An appeal process for the student speakers who believe the peer evaluations they received were inappropriate can be provided.

By using gists, advanced gists, and peer evaluations throughout the semester, critical listening becomes an integral part of the class. Teachers must realize, however, that documenting increases in critical listening is difficult. ESL researchers Kitao and Kitao note that "as listening proficiency gets more advanced, testing listening becomes more complicated." These researchers maintain that as listeners become more skilled, it "becomes more difficult to separate listening from other skills" and that testing methods often "do not reflect real-world listening tasks" (1996, Summary, para. 1). On the positive side, teachers should be able to observe growth in student facility in writing advanced gists and peer evaluations. Common sense would suggest that practicing these skills regularly would make students aware of critical listening and help them develop this valuable skill.

CONCLUSION

We need to heed the expert advice that tells us that speaking and listening skills are vital to student success. We can support the inclusion of speech courses in our general education curricula, but we can do more. We can, without adding an excessive burden to our teaching, include speaking and listening skills as an integral, valued part of technical-communication courses. Not only *can* we do this, we owe it to our students to do it.

APPENDIX A
The Pitch-Proposal Assignment

Introduction

This assignment was designed for a technical-writing class in which the redesign of a web site was the core project, but it can easily be adjusted to accommodate other projects.

Assignment

Your goal in the Pitch Proposal (PP) is to present a WWW site that needs redesign and convince classmates to join your design team. WWW sites that are personal home pages are not appropriate subjects for site redesigns. Sites chosen for redesign should be of appropriate taste for classroom presentation and discussion. If you have any questions about whether the site you are considering for redesign is appropriate, please e-mail the site URL to me (ericsson@wsu.edu) for consideration well before your presentation.

The PP has an oral and a written portion and concludes with the formation of teams to redesign the chosen sites.

Oral Presentation

The oral presentation must be from three to five minutes long. It must be a persuasive presentation that provides:

- An introduction to and review of the site you are suggesting for redesign
- A critique of site, pointing out the need for redesign
- A rough overview of your redesign plan
- Your personal expertise and the expertise needed from teammates
- An overview of what team members will gain from working on this project.

The goal of this presentation is to persuade class members to join you in the redesign project. It must be delivered extemporaneously from a key-word outline. It cannot be read from a manuscript.

Written Presentation

The written proposal for the PP will be a brief proposal written as an unsolicited proposal (an internal proposal designed for use within a company environment). For our purposes, the classroom is the environment. This proposal must include

- A problem statement
- An objectives statement
- A methods overview (focusing on personnel and schedule)
- Expected results

Graphics are optional. This proposal is due the day you give your oral pitch.

APPENDIX B
The Difference between Oral and Written Presentations
(in-class exercise)

Introduction

This in-class exercise is designed to make students more aware of the differences between these two modes of presentation. Following the explanation, there are exercises that students complete in teams of two.

Assignment

Has this ever happened to you? You are at a lecture or other presentation, and the speaker is reading from a prepared manuscript. As hard as you try, you can't follow what the person is saying. You try even harder and find yourself more confused. Finally, you give up. You may blame yourself for this experience by assuming that you haven't paid close enough attention, that you're not smart enough to understand the content, or that you should have gone to bed earlier the previous night. But the fault might not be yours. The speaker may have prepared a manuscript like he or she composes a text that is meant to be read, not heard. As smart as the speaker might be, he or she might not realize that there is a big difference in the way that we process written and oral language. As you are preparing to deliver your oral Pitch Proposal, you need to be aware of these differences.

Research on how we understand and process oral vs. written communication has discovered that readers and listeners need different cues to process information. For example, as listeners, we need explicit "signposts" that are repeated frequently so that we can track an oral presentation. Listeners don't have the benefit of punctuation marks that signal things like the end of a sentence or the beginning of a paragraph. When we read, we can back up and reread if we get lost, but when we're listening, that's impossible. As readers, we need a whole collection of indicators of how we should interpret a text—indicators that speakers can supply through inflection, volume changes, hand gestures, or facial expressions.

Considering linguistics scholar M. A. K. Halliday's research can help us understand some of these differences more easily. Halliday's research (1989) shows that written expressions have more "lexical" or content words than spoken expressions, whereas spoken expressions have more grammar/function items. For examples:

A *wagging, eager dog* met *Melissa* at the *open gate*. (7 content items—in italics; 3 grammar/function items)	*Melissa opened* the *gate* and as she did so she was *met* by a *wagging, eager dog*. (7 content items—in italics; 9 grammar/function items)
A *diverse music collection illustrates interest* in an *eclectic strategy*. (7 content items—in italics; 3 grammar/function items)	If you are *interested* in an *eclectic strategy*, you are willing to *collect* a *library of diverse music*. (7 content items —in italics; 11 grammar/function items)

Figure 1. Content/function.

Halliday calls these differences a matter of density. When compared to each other, "written language is dense, spoken language is sparse" (pp. 61-62). When you deliver a presentation extemporaneously you are much more likely to automatically use language that has more grammar/function items. That's good news for this class, because you're required to deliver your presentations that way. But it's good to experiment and realize the differences between written and oral language. Even an oral presentation that's read from a manuscript can be adequate IF it's been composed with the differences between written and oral delivery in mind.

Exercise #1

The escalating cutbacks in aerospace, defense, and other goods-producing industries have narrowed career opportunities for many of today's science and engineering graduates.

How many content items are in this sentence? How many grammar/function items?

Rewrite the above sentence for oral delivery. How does the ratio of content vs. grammar functions change?

Exercise #2

Motivation is essential for marketing work. Energetic professionals who function with minimal supervision are prized.

What is the content item count? What is the grammar/function count?

Rewrite the selection for oral delivery and count the same items.

Exercise #3

For a final example, take a sentence or two from the draft of your Pitch Proposal (the written one) and count the content vs. grammar/function items. Then rewrite the sample so that it is more appropriate for oral delivery. How does the ratio change?

APPENDIX C
Effective Use of Your Voice (in-class exercise)

Introduction

This in-class exercise is designed to make students aware of ways their voices can be used to enhance oral presentations.[5] *After being split into groups of 5 or 6, students practice delivering the "garbage" sentence 15 ways. Each group selects one member to deliver the sentence in each way listed. Each group member is required to deliver the sentence at least one way. This exercise provides considerable entertainment, while emphasizing the different ways the voice can be used.*

Assignment

Your practice sentence is "Hey, you sure do know a lot about garbage."

1. Use a monotone; don't vary the pace.

2. Make graduated changes in volume.

3. Say it very softly.

4. Say it in a soft, breathy voice.

5. Say it soft and sweetly.

6. Say it very loudly.

7. Say it in a loud announcer's voice.

8. Say it as quickly as you can.

9. Say it angrily, but softly.

10. Say it angrily and loudly.

11. Say it as slowly as you can.

12. Use pauses between every few words.

13. Stress one word. (Vary that word for different effects.)

14. Change the pitch to sound tentative.

15. Change the pitch and volume to sound positive.

---

[5] Thanks to Dr. Patty Sotirin of Michigan Technological University for introducing me to this exercise and providing the original materials for it.

APPENDIX D
Advanced-Gist Listening Assignment

Before starting to write advanced gists, students will have written and received feedback on basic gists. The advanced gist requires that students evaluate the quality of the supporting details. I borrow from a speech text I have used frequently, Osborn and Osborn's *Public Speaking* (2000), to form the critical listening questions advanced gists emphasize. Using all eleven of the questions from *Public Speaking* would be too complicated for a single advanced gist, so just two or three questions are used in each assignment. Teachers will notice that some questions are more appropriate for technical-communication presentations than others. I often reword the questions, but stating them in Osborn and Osborn's words is a good place to start.

Questions for Developing Critical-Listening Skills

1. Does the speaker support ideas or claims with facts and figures, testimony, and examples or narrative.
2. Does the speaker use supporting materials that are relevant, representative, recent, and reliable?
3. Does the speaker cite credible sources?
4. Does the speaker clearly distinguish among facts, inferences, and opinions?
5. Does the speaker use language that is concrete and understandable or purposely vague?
6. Does the speaker ask you to ignore reason?
7. Does the speaker rely too much on facts and figures?
8. Does the speaker use plausible reasoning?
9. Does the message promise too much?
10. Does the message fit with what I already know?
11. What other perspectives might there be on this issue? (pp. 72-75).

To avoid simplistic "yes" and "no" answers, I provide students with an example of an advanced gist that has received a ✓+ grade. Advanced gists should be limited in length, since evaluating 20–25 advanced gists, each several pages in length, would be too time-consuming. With a bit of experience, teachers can determine which questions are most appropriate and how long a good response might be. Setting a minimum of one page, double-spaced, and a maximum of two pages usually works well.

APPENDIX E
Advanced-Gist Listening Assignment

Introduction

These evaluations focus on oral delivery skills that have been emphasized in class. It's more effective to limit the evaluation to a few main skills rather than to ask the evaluators to evaluate the entirety of the presentation, especially since this evaluation form requires written and numerical ratings. Insisting on discursive and numeric evaluations gives the speaker a richer, more nuanced evaluation.

Assignment

You are required to rate and write an evaluation for one of your class members. While you are listening, rate the speaker on the following handout. Also take very brief notes so that you can write an evaluation later. You must hand in one copy to me and give one copy to the speaker.

Your written evaluation should be in prose/paragraph form, not in a numerical chart. You might write something like

> I thought your introduction was interesting because in your very first sentence you asked a question that I wanted to know the answer to. That was a good way of getting my attention and "hooking" me for the rest of the speech. I rated you 5 on the interest element. You had a theme too. The use of "the bigger, the better" made sense with your focus, and introducing it right away was great. I rated you a 5 on the theme too. I gave you a lower rating (4) on the thesis because I didn't really understand what point you wanted to make at the beginning of the speech. A clearer thesis would have made your speech easier to follow.

You would continue commenting on each point in the evaluation scale.

The final evaluation paragraph should include an overall commentary on the success of the speech and any pointers you'd like to give the person for improving his or her next speech.

You can use the following handout for rating and your notes:

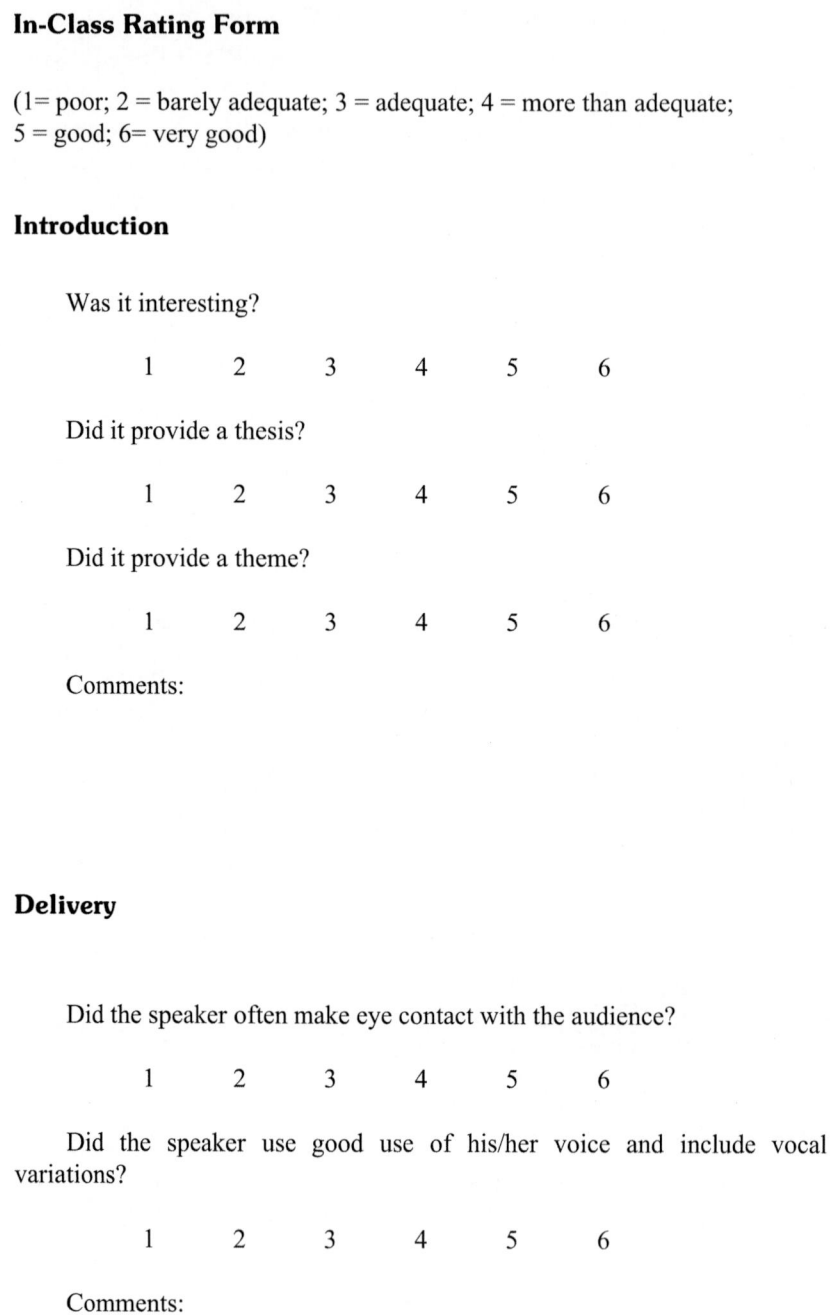

In-Class Rating Form

(1= poor; 2 = barely adequate; 3 = adequate; 4 = more than adequate; 5 = good; 6= very good)

Introduction

Was it interesting?

      1   2   3   4   5   6

Did it provide a thesis?

      1   2   3   4   5   6

Did it provide a theme?

      1   2   3   4   5   6

Comments:

Delivery

Did the speaker often make eye contact with the audience?

      1   2   3   4   5   6

Did the speaker use good use of his/her voice and include vocal variations?

      1   2   3   4   5   6

Comments:

Figure 2. In-Class Rating Form.

REFERENCES

American Express Small Business Services. (2000). *Voices from Main Street: Assessing the state of small business workforce skills*. New York: American Express.

Bajaj, G. (2003). Powerpoint repurposing. *INDEZINE.* fwww.indezine.com/stuff/pptrepurpose.pdf. Retrieved June 28, 2004.

Barthel, A. (2001). *Academically speaking: New identities or old realities.* Changing Identities Language and Academic Skills Conference. http://learning.uow.edu.au/LAS2001/unrefereed/barthel.pdf. Retrieved June 28, 2004.

The Boyer Commission on Educating Undergraduates at the Research University. (1998). *Reinventing undergraduate education: A blueprint for America's research universities.* http://naples.cc.sunysb.edu/Pres/boyer.nsf/. Retrieved June 26, 2004.

The Boyer Commission. (2001). *Reinventing undergraduate education: Three years after the Boyer Report.* New York: The State University of New York.

Halliday, M. A. K. (1989). *Spoken and written language*. Victoria, AU: Deakin University Press.

Hansen, R. S., & Hansen, K. (2004). What do employers *really* want? Top skills and values employers seek from job-seekers. *Quintessential Careers.* http://www.quintcareers.com/job_skills_values.html. Retrieved June 20, 2004.

Kitao, K., & Kitao, K. (1996). Testing listening. *The International TESOL Journal.* http://iteslj.org/Articles/Kitao-TestingListening.html. Retrieved July 5, 2004.

The National Communication Association. (2003). *Communication in the general education curriculum: A critical necessity for the 21st century*. Washington, DC: National Communication Association.

Noll, C. L., & Wilkins, M. (2002). Critical skills of IS professionals: A model for curriculum development. *Journal of Information Technology Education.* jite.org/documents/Vol1/v1n3p143-154.pdf. Retrieved June 22, 2004.

Osborn, M., & Osborn, S. (2000). *Public speaking* (5th ed.). New York: Houghton Mifflin.

Petress, K. C. (1999). "Listening: A Vital Skill." *Journal of Instructional Psychology.* http://www.findarticles.com/p/articles/mi_m0FCG/is_4_26/ai_62980773. Retrieved April 25, 2005.

Purdy, M. (2004). *The listener wins.* Monster Featured reports. http://featuredreports.monster.com/listen/overview/. Retrieved June 22, 2004.

SCANS 2000. (2000). The Workforce Skills web site. Johns Hopkins University Institute for Policy Studies. http://www.scans.jhu.edu/NS/HTML/SkillsDet.htm#Basic. Retrieved June 22, 2004.

Schilling, D. (2004). Be an effective listener. Womens'Meida.com. http://www.womensmedia.com/seminar-listening.html. Retrieved Feb. 2, 2005.

CHAPTER 15

Let's Talk: Preparing Students for Speaking and Listening in the Workplace

Gary Bays

> **OUTCOME:** Students will be able to speak and listen in small, informal group settings and collaborate effectively when presenting material.

BACKGROUND

The red curtain opens to reveal an intimidating auditorium. A bored audience stares back at you. One person in the crowd seems to be falling asleep; another coughs loudly and stretches his neck. You notice that your palms are sweaty. Your stomach is fluttering. You wonder whether you will pass out.

This is no ordinary panic attack: it is a virtual scene that was created to help people overcome anxiety about public speaking (Lubell, 2004).

Most of us are familiar with the polls representing the fear of public speaking. *The Book of Lists* ranks it the number one fear of Americans, ahead of dying itself, which ranks sixth, evoking the Jerry Seinfeld routine about how people would rather be the subject of a eulogy rather than be the eulogist (Rolls, 1998; Samson, 2000). And in response to those findings, researchers, scholars, and physicians seek ways to deal with the fears. Those responses to speaking anxiety offered by self-help and collegiate texts cover the gamut: think of the audience naked, find a friendly face in the crowd, visualize a positive engagement. In cases of severe phobias, medical personnel are now testing, in conjunction with the virtual reality sessions, drugs that play upon brain proteins and promise to help patients unlearn particular fears. The treatment is similar to that used for post-traumatic stress disorders (PTSD) suffered by survivors of the battlefield and, more recently, the attack on the World Trade Center ("Virtual Reality Exposure," 2004).

There's no denying that public speaking can be traumatic for select members of the population. Yet, as the *New York Times* and others call attention to virtual reality and drug therapy, the question arises, should we be treating this fear *after the fact*? Should Americans assume virtual reality and pharmaceutical companies will deliver them from the evils of public speaking?

Rather than endorsing *ex post facto* strategies for overcoming anxieties related to public speaking, we should be increasing the number of *informal speaking opportunities* for our students each semester. Like the other skills we readily promote in our classes—writing, editing, analysis, design—speaking should be a routine part of the technical-communication curriculum, not an onerous task too often relegated to the end of the semester with scant opportunity for rehearsal and high stakes. If students have the chance to work on informal interpersonal speaking skills—in meetings, in collaborative writing sessions, in brief oral presentations—they learn the skills demanded by the workplace, and both industry and academic sources agree that speaking well and working collaboratively are vital skills in the twenty-first century workplace.

WHAT BUSINESS AND INDUSTRY TELL US

When our students join the workforce, they must be able to work in teams and give small-scale presentations to their supervisors and peers. **Their colleagues will not expect them to provide a state-of-the-company address. Instead, they will expect new hires to provide them with state-of-the-project updates.** According to Norm Blackwell (2004), Director of Employee Relations for the American Regions of Molex, Inc., a Fortune 500 company, "Everyone wants to hire folks who can express their ideas well. And most bosses look for one key: what can this employee do for the company and for me as I try to get my job done?" Blackwell notes most young employees can't bring that much to the table early in their careers. Thus, their opportunities to speak often involve conveying pertinent information to other employees. "For example, they might be asked to explain in a meeting how a new law impacts the company. It's not that middle managers can't read the law themselves, but they want the new employees to understand it fully. Making young employees present information in this way demands that they learn it first."

A quick look at many of the current technical-communication texts on the market make little mention of the true contexts for most speaking situations new hires face. Instead, those texts focus primarily on presentation skills: how to organize a speech, how to eliminate mannerisms, how to analyze an audience—all skills college graduates should learn in an introductory speech course. **"What they need to learn is how to inform and persuade others in teams. That's where all companies see the need for clear speaking,"** notes Blackwell. He adds, "Anyone who has had training in collaborative work will be at a distinct advantage when they go on the job market."

Blackwell is not alone in his suggestions to students. Studies link public-speaking skills to "measures of success" such as obtaining employment and advancing on the job (Fordham & Gabbin, 1996). And for those who eventually advance in their careers, speaking becomes all the more important. A recent survey of professionals in a variety of industries shows that employees rank their bosses "average to poor when it comes to communication. A whopping 91% of those surveyed said effective communication is 'critical' to leadership, yet less than a third said that there were enough true voices of leadership in their organizations." The survey also revealed that 49% claim "their boss is average or weak when communicating at a human level. . . . **Running meetings and sharing critical information with employees are essential day-to-day activities for a leader**" ("New Survey," 2004).

In a survey of engineers with an average of 15 years work experience, responses revealed "meetings were the most important type of oral communication event." As one engineer noted, "'Formal presentations are few but on a weekly basis I am heading meetings that require speaking in front of small groups'" (Darling & Dannels, 2003, p. 8). The same survey found that "skills like teamwork, negotiation, and asking and responding to questions" rank high as attributes expected of employees. Perhaps most telling, "Seventy-two percent of the responses indicated that speaking skills (i.e., audience analysis, interpersonal communication, persuasion, confidence, teamwork) were important to engineering work." As the authors conclude, **"Engineers communicate interpersonally, in small groups, and on teams almost daily. The most important communication skills to develop are those related to oral performance"** (Darling & Dannels, 2003, p. 12).

WHAT ACADEMIC RESEARCH TELLS US

Students should be presented with a range of strategies for communicating information orally within small groups; they should be given ample opportunities to practice their informal, small-group speaking and listening skills with one another while in our classes; and we should find more effective ways to assess and improve these skills in more of the technical-communications assignments we give. As the work of Johnson, Johnson, and Smith (1991) shows, "Individuals, working together, construct shared understanding and knowledge" (p. 1:11). Their work also reveals that when students work cooperatively they "are less biased and have fewer mis-perceptions in comprehending the viewpoints and positions of others" (p. 2:6). The latter is important considering the discriminatory practices that reveal themselves when peers interact verbally in the workplace—practices that "dictate salaries, promotion, and power" (Cho, 2004). As Hamm and Adams (1992) point out, "Studies show that interaction among peers causes students—especially those from diverse cultural and linguistic backgrounds—to make significant academic gains compared to students in traditional settings" (p. 4).

Others go further: in his review of research on undergraduate education, Alexander Astin (1993) reported that "The single-most powerful force of influence on the undergraduate student's academic and personal development is the peer group" (p. 7). And as Kenneth Bruffee (1993) argued in his groundbreaking work on collaboration, critics are quick to fault American higher education for promoting passive, irresponsible, authority driven, and overly competitive undergraduates. Group work, particularly speaking in groups, is one means of countering those traits. **Indeed, teamwork and group presentations prepare students for the social, cultural, and professional demands awaiting them.**

Despite the merits of working and speaking in groups, that work brings with it some natural opposition and peril. The advance of widespread communication technology has made speaking less valued by the students themselves. Merrier and Dirks (1997) report that students "perceive oral communication less positively than written or E-mail communication." Finally, granting peers authority, accepting authority, and concerns about fair grading come with the territory. As Bruffee reminds us, "Social engagement can be hard work" (1993, p. 34). There's no denying that concerns about systematically assessing students' individual performances and grading fairly on final collaborative projects come with the territory when assigning group work, and that may take teachers out of their comfort level. The "hard work" Bruffee describes is not relegated solely to students.

If we are to make oral communication a staple of our classrooms, we must design courses that provide plenty of time for student interaction. "Students need time, long stretches, and regular ongoing opportunities to explore and understand difference and begin to find ways of talking and writing that encourages mutuality and reciprocity" (Fox, 1994, p. 119). Michaelsen, Knight, and Fink (2002) call for restructuring courses that put teams and a logical sequence of tasks at the center of instruction and learning.

UNDERSTANDING PROFESSIONAL PRESENTATIONS

Before we can design more opportunities for informal group work that focuses on oral communication projects and that introduce students to the kinds of daily workplace practices they will encounter after graduation, we must first put their speaking in context. **As the work of Darling and Dannels (2003) indicates, formal presentations are not the daily standard, nor the basis of evaluation, for most employees.** The fear of public speaking shared by so many American adults may be rooted in the perceptions and misconceptions we hold of public speaking itself. Too often we compare our speaking abilities to professionals who are buoyed by an array of writers, coaches, and technologies.

For example, when we view the State of the Union address each winter, we watch the president stand before a room full of politicians—many of them skilled speakers themselves—as well as a nationwide audience. The commander in chief seemingly speaks from memory and turns one well-oiled phrase after another. Unfortunately, many of our students are unaware that the president has a handful of speechwriters crafting his addresses; he speaks from a teleprompter that gives the illusion of eye contact and memorization; he also has media specialists choreograph his movements; and perhaps most important, politicians speak in public frequently—they have opportunities to practice the craft and speak to varied audiences throughout the day.

We need to discuss these elements of public speaking with our students at the outset of the semester so they have realistic expectations for the types of speaking they will do as well as the standards by which they will be judged.

CLASS PROJECTS

The following speaking opportunities build on the research and expertise cited thus far in the chapter. The principal guidelines are as follows:

- Students meet regularly in groups and use informal oral communication strategies to conduct meetings with specific outcomes.
- Each student takes on a leadership role across the course of the semester, using informal oral communication strategies to help a group be effective and efficient in completing an assigned task.
- All meetings require minutes and agendas to document what was accomplished in group work and to capture a sense of how effective students are in communicating orally with their peers.

All documentation from meetings is turned in at the subsequent class session so group members have a chance to review and amend their work if necessary. The minutes also allow me the opportunity to see how the group is functioning and, in particular, if they are on task. While I prefer not to involve myself in the group work outside of observing them in session and providing parameters for the work that is assigned, I want to ensure that groups are also not squandering their time together. To date, the practice has affected productivity; **students recognize their work is being assessed after each session, and they realize they are accountable for running efficient meetings with specific outcomes.**

As an introductory strategy, I introduce students to group work through a low-stakes exercise. Because students have strong convictions on the music they listen to, it becomes a natural topic for an informal speaking opportunity. Initially, students provide me with their top three favorite forms of music; this allows them to self-select into groups of similar interests. In some genres—hip-hop and country—there may be more than one group representing a particular style, but students quickly establish camaraderie with classmates holding similar tastes. Groups are limited in size to five so they become functional.

In short order, students select an artist or group they would like to bring to campus the next semester. Each student comes to an initial meeting with an artist proposal and a brief, five-minute rationale for their selection. That rationale, or pitch, should include

- Why that artist is likely to sell tickets on campus, as well as locally and regionally;

- Why the artist is more likely to draw interest than *other genres* or, potentially, other artists in the *same genre* (if there is more than one group working in that genre);
- When a performance date would best suit the school and the performer's schedules.

This initial meeting demands that students draw on their rudimentary rhetorical skills to argue their case. It also demands that each student visit web sites to seek information about artist profiles and tour dates. This task puts students in their comfort zone. Students are required to bring an informal one-page outline from which to speak and to share with group members. In the rare instance when two group members pitch the same artist, I simply ask that students consider the proposals objectively.

When the individual proposals are on the table, the group evaluates the artist proposals using a simple rubric that asks about detail and strength of argument. Students collect the responses from one another at the end of the class session and leave knowing how they performed during the exercise. I also use this in-class exercise as an opportunity to present and discuss Request for Proposals (RFPs) from the entertainment industry (see Appendix B), allowing students to review the features and commonalities in these documents.

Before going any further, we discuss the nature of the collaborative work mentioned earlier in this chapter. Very quickly they identify the major issues: division of labor, peer review, opportunities for negotiation, demands of leadership, dealing with freeloaders, and fair grading.

Having set a tone for regular group work in class, the formal group-speaking project asks that students compose and present a proposal for campus renovation. Because my college has been awaiting the release of state funds for a new student center, I've asked students to propose what that center should be (any current campus project serves as a potential topic, particularly if students have a vested interest in it (see Appendix A). They understand that their proposal will be presented to the decision makers on campus who are themselves designing the new student area (members of the administration, the business manager, the facilities manager). Intrinsically, students also see this as a way to voice their opinion on the future of the institution. The latter has been a strong motivator and has provided students with a very real audience.

Their first task is to compose an e-mail to me and to group members identifying three problems with the current student center and three suggestions for solving them. When there is overlap on suggestions, as often happens regarding lighting, sound, and décor, group members note it. Later this overlap is a way of reaching consensus when they organize their final presentation.

Chairs for subsequent meetings are chosen on a rotating basis. With each ensuing meeting, the groups discuss prompts from me and craft their proposal, dealing with traditional report components: rationale and significance, solution, timeline, personnel, graphics. **Each session prompts the production of text, but the vital**

outcome is that students speak with one another on a weekly basis as they grapple with the complexities of the project. Since technical communication on my campus is a survey course, my classes contain a literal range from future accountants to zoologists. Obviously, I don't require students to cost out the project or to acquire detailed bids for the work; nonetheless, students increasingly impress me with their knowledge of architectural software, CAD, spreadsheets, and digital photography that routinely augment their work.

As Michaelsen and others (2002) have discussed, the reliance on group work requires a time commitment from instructors. Across the semester, I devote *6–8 full class sessions* to group work, and portions of another handful. Particularly on a commuter campus where students work varied schedules, I cannot expect classmates to meet outside the class. **Repeated opportunities to practice communication skills are valuable to students, and by redesigning the course and response to readings, I don't feel I've sacrificed course content.** Indeed, I now cover meeting etiquette and efficiency (see Ann Marie Sabath, 1993), collaborative writing strategies and software, and group research and editing—material that was not a part of my class five years ago.

By the end of the semester, students are more comfortable speaking with and for one another, just as they will need to be when they enter the world of work. Each group presents their final proposal to the class, and they recognize that members of the administration will see the final draft of their proposal. In addition, they also evaluate one another's contributions to the project in an e-mail to me. The latter is intended to hold "freeloaders"—a concern of many students—accountable to their peers. Both of these final evaluations represent a percentage of the project grade.

CONCLUSIONS

Certainly, the documents that students produce collaboratively are markedly better than individual proposals I used to review, and **in a world that celebrates the *promise* of virtual reality, the class sessions promote the real *reality* of today's workplace.** Perhaps most important, as Thia Wolf suggests, "If the university is the last rite of passage into the adult world for some, into the professional world for others, we must provide our students with models of conversation, critical reflection, and action that they can use in the many institutional settings they will inhabit next" (1994, p. 109).

APPENDIX A
Collaborative Proposal: New Student Center

Wayne College invites design proposals for the new student center identified as a priority in the institution's master plan (copies of the master plan are available in the Wayne College Library). The new center will increase the current student area by 2,500 square feet.

Wayne College requests that all proposals address the following:

- Space for student-activity performances
- Ample room for dining and studying
- Design features that match the institution's character and decor
- Compliance with ADA guidelines
- Good circulation and traffic patterns
- Ease of maintenance
- Facilities for student organizations
- Room for continued expansion and increased student population

Proposals should follow the format discussed in class: title page, executive summary, table of contents, rationale and significance, plan of work, personnel, timeline. All graphics should be referred to in the text and aptly titled. Design, format, and paginate the document as discussed this semester.

You will receive a single grade for the document; be sure all group members participate fully. Proposals are due **December 5, 2006, at 5 p.m**. No late submissions will be accepted.

APPENDIX B
Requests for Proposals in the Music and Entertainment Industry

SUNDANCE INSTITUTE

Sundance Documentary Fund
Proposal Checklist for Development Funds

Date:

Applicant Name:

Title of Project:

Applicant Contact Information:

Address:

Phone: Fax: Email:

Brief project summary:

Proposal Checklist:

1. Project treatment/synopsis in English   ☐
   - a. Statement of objective   ☐
   - b. Narrative summary   ☐
   - c. Status of project   ☐
   - d. Director filmography   ☐
   - e. List of project's key personnel   ☐
   - f. Filmographies of key personnel   ☐
   - g. Distribution/marketing strategy   ☐
2. Comprehensive budget in U.S. $ ☐
   - a. Amount of funds secured to date   ☐
   - b. Complete list of funding sources   ☐
3. Director's completed prior work on VHS (NTSC, PAL or SECAM)   ☐
   with English subtitles or accompanied by an English dialogue transcript
   (If no video sample is available, a creative work indicating the director's
   artistic point of view and ability to tell a story using a visual medium,
   is required)

Label all tapes with proposal's title, director and producer's names, and tape's total
running time. Please note: Tapes will not be returned. Please DO NOT bind or staple
proposals. Please send proposals to: Sundance Institute, Sundance Documentary Fund, 8857 W.
Olympic Blvd., Beverly Hills, CA 90211 USA

THE GRAMMY° FOUNDATION

GRAMMY FOUNDATION RESEARCH GRANT PROGRAM

Please follow the Guidelines and Application instructions.
Applicants are required to submit one signed, original completed application.

NARRATIVE
(Number each narrative item as listed below. Must not exceed three single-sided sheets.)

1. Describe the project for which you are requesting funds.
2. State research question and include specific outline of methodology, outcome variables.
3. List key personnel involved in the project - include title or position.

BUDGET
- Use the budget format provided. *(No other format will be accepted)*
- Expenses paid by the GRAMMY Foundation grant funds must be clearly identified on the budget form using **ASTERISKS.**
- List all financial and in-kind partners.
- Adjust line items as needed.
- Indirect costs: The GRAMMY Foundation will pay up to 15% of indirect costs with supporting documentation, (i.e., letter from Human Resource department documenting this cost).

SUPPORT MATERIALS
1. Two letters of support for the project. These letters **may not** be written by the organization requesting funds or by individuals related to the project. General letters of support for your organization or letters supporting a different project do not qualify.
2. Letter of commitment from partnering organizations.
3. Organizations must submit general description and history of organization. (1 page only)
4. Individuals, Project Directors and/or Key Personnel must submit a brief, current biography. (1 page only)
5. Organizational income and expense statement (actuals) for last completed fiscal year
6. IRS tax determination letter for non-profit organizations

TO RECEIVE NOTIFICATION THAT YOUR APPLICATION WAS RECEIVED,
INCLUDE A SELF-ADDRESSED STAMPED POSTCARD.

MAIL COMPLETED APPLICATION TO:
The GRAMMY Foundation Grant Program
3402 Pico Boulevard
Santa Monica, CA 90405

REFERENCES

Astin, A. (1993). *What matters in college? Four critical years revisited.* San Francisco, CA: Jossey-Bass.

Blackwell, N. Personal interview, August 15, 2004.

Bruffee, K. (1993). *Collaborative learning: Higher education, interdependence, and the authority of knowledge.* Baltimore, MD: Johns Hopkins.

Cho, J. (2004, July 11). Workplace speech divided by gender; Communication styles differ widely. *The Times-Picayune* (New Orleans). Retrieved from LexisNexis, July 13, 2004.

Darling, A. L., & Dannels, D. P. (2003). Practicing engineers talk about the importance of talk: A report on the role of oral communication in the workplace. *Communication Education, 52*(1), 1-16.

Fordham, D. R., & Gabbin, A. L. (1996). Skills versus apprehension: Empirical evidence on oral communication. *Business Communication Quarterly, 59*(3), 88-98.

Fox, T. (1994). Race and gender in collaborative learning. In S. B. Reagan, T. Fox, & D. Bliech (Eds.), *Writing with: New directions in collaborative teaching, learning, and research.* Albany, NY: SUNY Press.

Hamm, M., & Adams, D. (1992). *The collaborative dimensions of learning.* Norwoord, NJ: Ablex.

Johnson, D. W., Johnson, R. T., & Smith, K. A. (1991). *Active learning: Cooperation in the college classroom.* Edina, MN: Interaction Book Company.

Lubell, S. (2004, February 19). On the therapist's couch, a jolt of virtual reality. *The New York Times.* Retrieved from LexisNexis, July 1, 2004.

Merrier, P. A., & Dirks, R. (1997). Student attitudes toward written, oral, and E-mail communication. *Business Communication Quarterly, 60*(2), 89-100.

Michaelsen, L. K., Knight, A. B., & Fink, L. D. (2002). *Team-based learning: A transformative use of small groups.* Westport, CT: Praeger.

New survey from Bates Communications reveals employees' dissatisfaction with their bosses' communication skills. *PR Newswire,* Retrieved from LexisNexis, July 13, 2004.

Rolls, J. A. (1998). Facing the fears associated with professional speaking. *Business Communication Quarterly, 61*(2), 103-106.

Sabath, A. M. (1993). *Business etiquette in brief: The competitive edge for today's business professional.* Avon, MA: Adams Media.

Samson, T. (2000, March 13). Master but don't fear presentations—Although some may feel that public speaking is akin to a fate worse than death, there are easy ways to rise to the occasion. *Infoworld, 22*(11), 5.

Virtual reality exposure cures patients' fears. (2004, March 21*). Hospital Business Week.* Retrieved from LexisNexis, July 1, 2004.

Wolf, T. (1994). Conflict as opportunity in collaborative praxis. In S. B. Reagan, T. Fox, & D. Bleich (Eds.), *Writing with: New directions in collaborative teaching, learning, and research.* Albany, NY: SUNY.

CHAPTER 16

The Elements of Editing: Relationships, Roles, and Revisions

Dànielle Nicole DeVoss

> **OUTCOME:** Students will produce better, more polished, more refined prose; understand the different levels of editing and approaches appropriate for each; and be familiar with the editing approaches appropriate for different types of writing and documents.

> Writing skills are fundamental in business. It's increasingly important to be able to convey content in a tight, logical, direct manner, particularly in a fast-paced technological environment.
>
> (College Board, 2004, p. 10)

WHAT BUSINESS AND INDUSTRY TELL US

Most workers with communication tasks will engage in editing. As *Writing: A Ticket to Work . . .*—a report providing data from a survey of industry leaders representing companies employing almost eight million workers—notes, two-thirds of employees in North America have some writing responsibility (College Board, 2004). Being expected to write means that writing tasks are part of a job description. *All* employees perform writing at work and are judged and sometimes reviewed on the effectiveness of their writing. Editing is a part of all communication tasks—from quickly polishing an e-mail before sending it to a team or a supervisor to thoroughly reviewing a longer document before delivering it for printing.

Job posts on any of the large, national job banks reflect the importance of effective communication. Companies seek candidates who can develop, research, manage, and produce effective communication for audiences both within and outside of the organization. **Part of producing accurate, informed, persuasive prose is being able to edit one's own written work, and often the work of others.**

The ability to produce clear, correct prose is a skill that most businesses seek in employees. Reports from industry reveal that writing is one of the most important skills employers seek, yet often a skill lacking in most job applicants (see, for instance, Gaidosch, 2003; Grensing, 1988). A survey conducted in 2004 by the College Board's National Commission on Writing reported that most U.S. employers feel that "about one-third of workers do not meet the writing requirements of their positions . . ." and that "in a fast-paced workplace, precision and brevity are essential" writing skills required of all employees. The survey found that the most sought-after writing-related skills were "accuracy, clarity, spelling, punctuation, grammar and conciseness" (Read, 2004, online).

Digital tools and electronic spaces are changing the amount and types of writing workers produce, but conventional criteria of professionalism and correctness continue to be important. The biggest change in the ways in which writers communicate on the job is e-mail. Relatively unknown in the North American workplace in the early-to-mid 1980s, e-mail is now the primary mode of communication: "Because of e-mail, more employees have to write more often" (College Board, 2004, p. 14). Although it is easy to think of e-mail as a less-formal space for writing—because of the ways in which e-mail is used to routinely deliver workplace documents and because of the ways in which it is more "conversational" than formal, print-based prose—e-mail messages must be as carefully edited and polished as other documents. Along with e-mail, "oral presentations with visual aids (e.g., PowerPoint)" have become "ubiquitous in the American economy" (College Board, 2004, p. 13).

The messages created with these new technologies, however, are still often evaluated with conventional criteria of professionalism and correctness. A recent query distributed on SmallBusinessComputing.com, "Do correct spelling and grammar matter anymore?," generated the following response:

> I do hesitate when finding these types of errors at a website when browsing as a buyer. Of course, I tend to excuse a major brand for that kind of error at its website, but am less patient when surfing a lesser known competitor.

> Online reputation is developed, in great part, on the accuracy of spelling, grammar and punctuation. At least, that is what I believe . . . (Windhaus, 2004).

Similarly, Charles Rubin (2005), in an article "You Are What You Write," notes that

> The words you write are your best and often only chance to create an impression on customers. . . . **Before you will ever have the chance to impress customers with your superior service, expertise, or follow-up, many of your prospects will decide whether or not to do business with you by the quality of your written communications.**

> None of us likes to think our writing skills need improving, and even fewer of us are willing to spend the time and energy to improve them. But consider this: your larger and better-financed competitors pay professional writers to

create their storefront copy, produce electronic publications and brochures, and perhaps even pen their e-mail for them. In order to compete, you must develop the same professionalism in your writing.

Editing is recognized as complex, multifaceted work on several levels. Editing one's own work is itself a significant task. Editing the work of co-workers adds a new layer of difficulty, especially when writing projects constitute part of a larger team effort. A recent Department of Labor report, for example, argued that the capabilities that employees need in today's workplace go beyond a set of "basic" individual skills; rather, they are anchored in an employee's ability to work as a member of a team—engaging in verbal, written, organizational, and interpersonal communication with others (Business-Higher Education Forum, 1999; U.S. Department of Labor, 1999).

Adding to the complexity of communication-related skills is the fact that editing means different tasks and has different definitions depending on where and when it takes place. Most often, editing refers to the close, careful review of text to verify a document's grammatical and mechanical correctness. Editing can also refer, however, to the deeper processes of revising a document, which may require a writer to significantly rearrange, rethink, reorganize, and redevelop pieces of a document or an entire document. Editing can also mean reviewing graphics and other images, tables and charts, and other document elements.

WHAT ACADEMIC RESEARCH TELLS US

Editing is a living, dynamic art. Editing evolves due to a variety of historical, cultural, technological, and rhetorical transformations—from changes related to downsizing to changes in new technical lexicon that come with new discoveries; from single-sourcing approaches to distributing documentation (Dragga & Gong, 1989; Gross, 1993; Plotnik, 1982; Rude, 2002; Samson, 1993). Editing tasks have shifted in today's global marketplace due to the ways in which many documents cross cultures and national boundaries (Leininger & Yuan, 1998).

Notions of editing and revising are typically different within academia and within workplaces. In college courses, editing, especially for technical documents, often relates to stylistic changes—checking grammar and proofreading mechanics. In industry, however, revision often means grand-scale changes launched by various parties involved in a document's creation and production, including multiple authors and, at times, technical editors assigned to a project. Revision might best be conceived of as "reconceptualizing, revisualizing, reevaluating, refocusing, reorganizing, rewording, and reformatting" (Gerich, 1994, p. 59; see also Haugen, 1991).

Editing in most large corporations is guided by in-house style guides. Most large companies and even some smaller organizations maintain and follow their own internal style guides. Writers must thus negotiate institutional stylistic choices. Style guides are useful tools because they not only guide often-subjective stylistic choices, but they also raise awareness of writing standards and the importance of maintaining such standards (Caruthers, 1986).

Digital tools and electronic spaces are changing the types of writing communicators prepare and edit as well as changing the ways editing is done. New media and new technologies have changed the scope of writing. Although editing visual media has always been a task of editors, the availability of graphic design and image manipulation software has broadened the types of material writers must negotiate. Editors must be able to make appropriate and defendable media choices—from assessing to creating tables, diagrams, charts, and schematics. This requires not only the ability to master the tools to construct such materials, but it also requires an ethical eye toward accurate and appropriate representation of data (Allen & Voss, 1998; Williams & Harkus, 1998).

One instance of writing and editing new media is reported in a study by Anderson, Campbell, Hindle, Price, and Scasny (1998), who studied a group of students working to edit a web site. The students found quite early on that the web site was not only an exercise in textual editing, but was also a task that required them to assess "screen design, coding, interface issues, and interactivity" (p. 47). Editorial shifts in attention and focus are needed when text moves from the page to the screen and when visual design and textual content collide.

Further, dynamic documents distributed via queried databases provide new types of editing situations. As Michael Albers (2000) described dynamic documents:

> Multiple writers at multiple locations contribute information to a document database which then, upon reader request, dynamically generates a unique document fulfilling current reader needs. What the reader sees is not a document that an editor has carefully groomed, but rather a dynamic document which was compiled from a database just before the information was presented.

Before easy access to robust word-processing software and other computer-based writing tools, much editing work was done on hard copy, by hand and red pen. Today's tools provide for a rich context of editing, from the basics (e.g., spelling- and grammar-checking functions) to more advanced document management features, such as the ability to control versions of documents and the flexibility to quickly prepare and share documents (Farkas & Poltrock, 1995). Editors thus work across hard-copy and electronic editing tools and practices (e.g., from the use of hard-copy markup before editing an electronic file to the use of the track-changes feature in word-processing software; Dayton, 2003, 2004a, 2004b).

TOPICS AND ASSIGNMENTS

Understanding the Roles of Editors and the Status of Documents

> The most difficult part of editing someone else's work is understanding their purpose and meaning . . . so that you edit in a way that remains true to their original intentions. It's very important to pry, probe, and ask questions in order to figure out what your role will ultimately be as an editor.
>
> (Angela, Professional Writing undergraduate student)

When helping students understand their roles in approaching an editing task, it is crucial for them to develop skills in selecting among the repertoire of rules and tools available. It is useful to discuss with students how to assess the context and needs of a particular document by, for instance, asking that students generate a set of questions they would use as a heuristic to frame their initial read and edit of a document. Questions that might emerge during this conversation include:

- What is the status of this document? Who has read it thus far? Who has helped in its writing? What significant changes have been made thus far?
- What sort of feedback is wanted on this document?
- What other edits do you plan while finishing the preparation of this document? When do you plan to do a next phase of revision and editing?

Understanding the Levels and Types of Editing

All writers who perform editing tasks need to understand the levels of editing, which typically fall on a continuum: developmental editing, comprehensive editing, specific editing, copyediting, proofreading. Discussing with students the different levels of edit is useful, as is assigning activities that allow students space to gain hands-on skill with different types of editing. Note that in the descriptions below, "editor" might refer to the author of the document, someone within a project team acting in the role of editor, or a writer with the formal title of technical editor. Note, also, that an ability to carefully, thoughtfully, and tactfully interact with writers while suggesting changes is a skill required across the levels and types of editing.

Developmental editing: An editor provides a global review of a document, providing suggestions that might include reorganization, deletion of entire chunks of the text, suggestions regarding additions that need to be made, etc. The editor's role is to help to develop the document. Tools needed: an ability to see the "big picture" and a strong sense of the history of the document and the future purposes and uses of the document.

Comprehensive editing: An editor provides a global review of a document, focusing on how well the organization flows and analyzing the placement of concepts, ideas, images, etc. The editor's role is to provide global comments on a document that has perhaps already been developmentally edited. Tools needed: an ability to see the "big picture" but to also focus on specific details. A strong sense of the history of the document and the future purposes and uses of the document.

Specific edit (e.g., on one aspect of a text or document): An editor provides feedback and suggestions regarding some aspect of the document; for instance, an editor might review all of the tables, charts, and diagrams; an editor might review the formatting and layout features to make sure they are on a par with the institution's internal style guides; an editor might carefully check and verify the facts and statistics presented in a document. The editor's role is to provide honed, specific editing related to an aspect of a document. Tools needed: an ability to see the "big picture" but to also focus on specific details. Familiarity with and access to the tools needed to assess specific aspects of a document (e.g., the ability to interpret and design tables); and access to the other writers or editors focusing on other aspects of the document.

Copyediting: An editor provides a focused, global review to assess a document at the sentence and paragraph level. When copyediting, an editor might focus specifically on flow, transitions, grammar, and style to make sure the document is as polished as possible, and that it is stylistically appropriate and grammatically correct. Tools needed: an ability to see the "big picture" but to also focus on very specific details. Attention to and tools related to assessing grammar, style, punctuation, and consistency across a document; an internal style guide, or sample internal documents to compare the document to; and an outside/general style guide (such as *Words into Type* or the *Chicago Manual of Style*).

Proofreading: This is typically a final phase of editing, where an editor reviews a document after the content is polished, finished, and reviewed, and after its final design and layout have been set, just before the document goes to press or is published on a web site, burned to a CD, etc. An editor provides an extremely focused, line-by-line review to check and make sure that no errors still exist and that no errors have been introduced during the preparation for publication. Tools needed: an ability to look past the "big picture" and to focus on very specific details, and the ability to compare a current version of a document to a prepreparation version so that any errors can be identified.

Introducing Activities and Exercises

The activities briefly described below are designed to help students:

- produce better, more polished, more refined prose;
- understand the different levels of editing, and approaches best applicable to those levels of editing;
- become familiar with the editing approaches best applicable to different types of writing and different types of documents; and
- successfully edit their own work and the work of others.

An exercise that lends itself well to all of the activities below is to ask students to reflect upon their approaches and processes as they engage any of the tasks—in a journal, on a blog, or via some other method. Being able to successfully edit is one skill, but being able to reflect upon and communicate editorial designs and processes is another skill entirely. Asking students to reflect upon the choices they have made and the tools they have negotiated helps to bridge both sets of skills.

DEVELOPMENTAL AND/OR COMPREHENSIVE EDITING

An obvious approach to large-scale editing is often put into place in writing classes already—asking students to bring in drafts of their work to share and to review. Reframing this activity with specific goals, prompts, or editorial advice to scaffold the peer-review process can make this an even more valuable activity. Not only do students receive thoughtful reviews of and suggestions for their work, but they have an opportunity to apply specific editorial approaches.

To help engage students in the processes shaping suggestions and communicating with other writers, students might be provided with a sample document and a prompt like this.

> A co-worker has shared with you a draft of a brochure he created for the nonprofit organization for which you both work. The brochure's audience is. . . . The brochure should help people understand. . . . Your co-worker knows that you have excellent writing skills and that you're good at providing feedback and suggestions. He asks you to review his work and asks for feedback and suggestions for making it better.

> Construct a brief (2-pages or so) memo to your co-worker making suggestions for the draft brochure he has written. You know you can't overwhelm him with a giant list of critiques and suggestions, so you decide to focus on a few (e.g., five) specific, important suggestions.

Another approach to developmental editing is to ask students to download a paper from an online, for-free paper mill (e.g., www.schoolsucks.com, www.cheathouse.com). Their first task might be to establish the assignment or task for which the paper was written (this information is typically not provided by the essay-download sites). After establishing a context, students might then engage in a developmental edit of the document. Online databases (especially those available through government sites), also offer drafts of reports, which lend themselves well to this activity.

Another approach to large-scale editing is to ask students to choose a technical article from a trade magazine or journal and translate it for a lay audience. This translation might be in the form of a PowerPoint presentation, a white paper, a short newsletter article, or a brief informational report. Students will have to make specific editorial decisions as they transition the prose from one audience to another (see Appendix A).

A somewhat similar activity is to select a print document and have students convert it for digital delivery. This will require that they think deeply about what portions can be cut, which need further development for the screen-reading context, what sort of navigation schemes are most appropriate for the material, etc. This assignment can range from developing a set of Web pages to parsing a document into pieces for database delivery.

Finally, consider asking students to download a house style guide or a corporate style guide and apply the principles in the guide to a sample document. Many sample guides are available online for download; two examples are

Microsoft's Manual of Style for Technical Publications
http://www.microsoft.com/downloads

Modern Humanities Research Association Style Guide
http://www.mhra.org.uk/Publications/Books/StyleGuide/StyleGuideV1.pdf

SPECIFIC EDITING

> The most difficult part of editing someone else's work is "understanding precisely their intentions and their style. It's easy to edit away from the point they want to make, and it's easy to make the work sound different."
>
> (Jon, Professional Writing undergraduate student)

To help students understand the wealth of information in almost any document, ask that they work from a sample document—such as a brochure for a local coffeehouse, a governmental fact sheet, or a company's annual report—and pull it apart. Have students work in small groups to establish lists of the different types of information included in each document, and to create a heuristic for editing the different types of information—for example, information contained in tables, information included in appendices (see Appendix B).

A project-based approach for engaging students in the processes of specific editing is to have students actually produce a document that includes different types of information, for example, textual information; statistical data in tables; graphics or other images; charts and diagrams; historical data; etc. After students have produced an initial draft of the document, establish editing teams to closely review specific content; for instance, one student group might be in charge of researching and verifying historical data or information, while another group might be responsible for verifying the consistency of design features across the document (e.g., heading levels, headers and footers).

COPYEDITING

> I hate (and love) the rigorous attention to detail that is necessary to edit. It's tiring and powerful.
>
> (Casey, English undergraduate student)

Starting with their own work is a good initial approach for engaging students in the task of copyediting: ask students to find and bring to class a paper they wrote at least 6 months prior. They might begin the task by refreshing themselves—for instance, by writing a 1-page memo reviewing the audience, context, purpose, creation date, and other details of the document—before copyediting the document and then adding an additional section to their memo, in which they reflect on what they've changed and what patterns that required correction were visible to them as they edited their document (see Appendix C).

If students are working on a large project, or working with a specific set of materials, ask that they create a copyediting manual. For instance, if a newsletter or journal is edited within the department or college in which the course is taught, students might review past issues, speak with the editor or editors, and work to establish an internal manual for copyediting the publication.

PROOFREADING

Cycles of proofreading can be integrated into a technical writing class by asking that students take on a careful copyediting of their own work and a thoughtful proofreading of each other's work before they hand in a final paper to an instructor. Students will have to negotiate the time they have and the time they will need to share their documents, review others' documents, and refine their own documents before an assignment is due. After the first instance of this activity, teachers might ask students to prepare a proofreading guidesheet—perhaps a one-page checklist of suggested errors to watch for.

Proofreading projects lend themselves well to partner projects—if an academic journal, student journal, or any newsletters are produced in a department or at an institution, collaborating with the editors-in-chief of such publications to make copies accessible to students across the writing and editing cycle is incredibly valuable. If this is not possible, at least making the documents available to students at the proof stage will provide students valuable experience with polished documents being reviewed prior to final publication.

Equipping Technical Editors with Tools and Resources

Along with the tools mentioned above (folded within the discussion of the levels of edit), good writers always have a variety of tools available to them, depending, of course, on the workspace and the preferences of the writer. Adding a set of editing tools will benefit any writer's toolbox.

One exercise for helping students to understand and use the editing-based tools available in word-processing software, specifically, is to have students work in small groups to prepare short recommendations for how editors can best make use of these tools and features. Each group might be tasked, for instance, with describing five ways in which word-processing software supports the work of editing. Each group can then plan to present their recommendations to the class, and the recommendations can be merged to provide a tool sheet for the entire class.

If time lends itself to extending the project outside of class, students might be asked to interview working technical writers to inquire as to what editing-based tools they rely upon on a regular or daily basis. Inviting these working editors to visit class and demonstrate the use of tools and how they apply the tools to sample documents lends itself well to helping students understand the application of different tools.

Other sample tools, most of which lend themselves to at least discussion, if not some in-class development and tailoring, include:

- a personal style guide, developed with time and across writing tasks (e.g., a notebook or electronic file in which an editor lists specific approaches, tips, or problems, for instance "always remember to look for the correct use of hyphens, en dashes, and em dashes" and "remember: it's company convention to never use vertical lines in a document")

- an editing portfolio, in which a writer keeps examples of edited documents, especially documents that provide a profile of writing approaches and particular stylistic rules

- an identified approach or rhythm for editing (e.g., copyediting in three sweeps: first, skimming the document to get the big picture; second, reading the document carefully for stylistic issues or organization concerns; third, reading the document carefully for grammar and mechanical problems)

- an approach for managing documents (e.g., retaining DRAFT on draft documents, including the date of last review within a document, naming a document so that an older version isn't saved over a new version—for instance GBR_report_v_2 7-12-04 and GBR_report_v_3 7-15-04)

- an approach to providing feedback in hard copy to other authors involved with a document or project (e.g., attaching sticky notes with comments; using a red pen to make marginal comments; attaching a cover memo to an edited document, with core areas of concern identified)

- an approach to providing feedback in electronic copy to other authors involved with a document or project (e.g., using the "note" feature of the word-processing application, using textual highlighting to insert comments, such as actually highlighting comments or marking comments [**INSIDE BRACKETS AND IN BOLDFACE TYPE**])

- an editorial library, including key reference books that the writer might need access to (e.g., a style guide, a grammar guide, any internal style guides, a dictionary, a thesaurus)

Often, a writer will not be told "we need a copyedit on this document." Further, authors may not know where their document is in terms of preparation and revision, or what sort of editing it might need. So writers might receive a document with a note attached asking to "review this," or a project team might ask them to edit a document. This makes the set of approaches and the tools listed above all the more helpful in approaching a task and engaging a document in the most appropriate level of edit for that particular writer, document, or project.

EVALUATING STUDENT WORK

Students should have not only the chance to perform editing tasks, but also the opportunity to reflect in rich and meaningful ways upon their editing work—to write about skills they learned, approaches they developed, and challenges they faced (and hopefully conquered). As an instructor, I find these reflective documents to be incredibly helpful in the evaluation phases of editing work.

Additional strategies for assessing students' editing efforts include the following:

- Because proofreading marks are shared and conventional, quizzes on such marks and the use of the marks can be helpful learning tools—tools that can also be graded and evaluated. Quizzes don't necessarily have to be mere memorization of editing marks, but can draw upon students' ability to identify, explain, interpret, and apply different marks across different types of editing contexts.

- Because editing is inherently collaborative and is often team driven, asking students to evaluate and reflect upon each other's performance—and to provide a written rationale for their assessment of each other—can be a useful tool for evaluation.

- Asking students to create sets of heuristics and rubrics can be incredibly useful in the classroom: students have to deeply contextualize their work, create connections across academic exercises and workplace conventions, and think about the assessment- and evaluation-related consequences of their work.

- Adopting a grading approach that draws upon the levels of edit can be applicable; for instance, students may receive points or a grade based on (1) their substantive or overall effort, (2) their work on specific areas within a document or project, and (3) their work on stylistic, mechanical, and grammatical aspects of a document or project.

CONCLUSION

This chapter addresses the complex and situated task that is editing technical writing. Situated within relationships, framed by various documents and different audiences, and often part of the daily work of writers in an organization or industry, editing is a complicated part of a communicator's job. This chapter offers a framework to better understand the ways in which technical editing is understood and undertaken both in workspaces and in academia and offers tools for students to become more familiar with the tools available to writers-as-editors. The suggested exercises provide a set of ideas for instructors to shape as appropriate in their own technical-writing classrooms, with the larger goal of helping familiarize students with the editing approaches best applicable to different types of writing and different types of documents.

APPENDIX A
Sample Assignment: Developmental Editing

Instructor Overview

Materials: This assignment requires that you select a set of technical documents to provide students. A honed Web search is an effective way to find a useful set of documents for this assignment. Journal articles are particularly useful documents to use for this assignment in upper-level classes in which students are relatively familiar with professional journals.

Evaluation: There are two products for this assignment: (1) the revised document that students produce, and (2) a 2-page memo in which students explain and justify some of the decisions made as they revised the document. The memo should be the guiding document, as it will reflect the thought and attentiveness that guided the revision of the document. Approaches to assessing and evaluating the revised document include:

- Is the justification and rationale for the revision descriptive, appropriate, and compelling?
- Are the main points accurately and appropriately translated and integrated into the revision?
- Does the entirety of the revised document support and elaborate upon the main points? Are the main points supported by understandable examples, good description, and appropriate details?

ASSIGNMENT 1: DEVELOPMENTAL EDITING

Setting the Stage

For this assignment, you will work in small groups to edit and prepare a highly technical document for a lay audience. This translation might be in the form of a PowerPoint presentation (slides and presentation notes), a white paper or brief informational report, or a newsletter article.

You will have to create a context for yourself and for the document; for instance, are you a field engineer presenting the findings of a survey to upper management, who may not be familiar with the jargon you use and the statistics you generate in your work? Are you a representative of a company presenting a recent land redevelopment project to a local community organization?

Getting Started

Read the document you have been provided. Decide upon an appropriate approach to revise the document for a lay audience. You might take into consideration aspects such as:

- the jargon (field-specific language) used in the document
- the length and prose style of the document
- the graphics or other visuals in the document
- the key/main points conveyed in the document

Next, decide what format you will use to deliver the document to a lay audience. Construct a list of the characteristics of that format (e.g., the typical design features of PowerPoint presentations, the typical main headings in an information report).

APPENDIX B
Sample Assignment: Specific Editing

Instructor Overview

Materials: This assignment requires that you first select a set of documents to provide students. You might either select a set of documents representing one genre (e.g., an annual report), or you might select a set of documents representing diverse genres. A honed Web search is an effective way to find a useful set of documents for this assignment.

Evaluation: There are two products for this assignment: (1) the revised document that students produce, and (2) a 2-page memo in which students explain and justify some of the decisions made as they revised the document. The memo should be the guiding document, as it will reflect the thought and attentiveness that guided the revision of the document. Approaches to assessing and evaluating the revised document include:

- Is the justification and rationale—the framework—for the lists appropriate and compelling?
- Are the lists thorough? Do they represent the multiple pieces of information included in the document?
- Are the heuristics thoughtful and appropriately complex? Do they serve as an appropriate set of guidelines for evaluating the information included in the document?

Setting the Stage

For this assignment, you will work in small groups to pull apart the document you are provided. Specifically, you will establish lists of the different types of information included in the document you are provided (e.g., photographic, statistical). Once you have established your lists, you will construct a set of heuristics for assessing and editing the different types of information represented in each list.

Getting Started

Read the document you have been provided. As you construct your lists, think about the ways in which you can categorize and label the information included in the document. As you work, you might consider aspects such as whether all of the "visual" content appears in one category. Or should figures, graphs, photographs, etc., be separated into different categories?

As you construct your editing heuristics, carefully consider the context and audience for the document. As you work, you might consider aspects such as how a photograph might be edited. What sorts of changes are appropriate to make to a photograph when preparing it to include in this type of document?

Producing and Presenting Your Lists and Heuristics

Each small group will present and explain their document, then introduce their lists and heuristics. In your presentation, be prepared to explain and justify the ways in which you have categorized the document content and the ways in which you have structured the editing heuristics.

APPENDIX C
Sample Assignment: Copyediting

Instructor Overview

Materials: This assignment requires that students select a document they wrote at least 6 months prior. You may also want to set specific requirements as to what type of document they select (e.g., the document genre, the document length, the context in which the document was produced). Students will also require a set of standard proofreader's marks. You can either construct this together in class, or you can provide these to students in the form of a handout. If you prefer that the students electronically proofread, you might discuss the "track changes" feature in the word-processing application available to you and encourage students to edit on screen.

Evaluation: There are two products for this assignment: 1) the copyedited document students produce, and 2) a brief (2-page) memo in which students explain the original context of the document and explain the copyediting changes made. The memo should be the guiding document, as it will reflect the thought and attentiveness that guided the revision of the document. Approaches to assessing and evaluating the copyedited include:

- Does the document reflect a thorough, careful copyedit?
- Does the document reflect the copyediting conventions discussed in class?
- Are there patterns of error that have not been marked?
- Has the author thoughtfully addressed and explained the copyediting decisions in the cover memo?

Getting Started

For this assignment, you will first select a document you produced at least 6 months ago. Then, applying the copyediting skills you have learned and honed in class, you will copyedit your document, using standard proofreader's marks.

Submitting Your Products

There are two products for this assignment: (1) the copyedited document, with clear application of standard proofreader's marks, or, if you have electronically edited, "track changes" on and visible; and (2) a 2-page cover memo in which you first describe the original purpose and goals of the document (e.g., audience, context, purpose, creation date, and other details of the document) and reflect upon the copyediting changes you made to the document (what changes did you make, specifically? What patterns of error did you notice in your own writing?).

ACKNOWLEDGMENTS

Thanks to the undergraduate Professional Writing and English students and to the graduate Rhetoric & Writing students who responded to my queries regarding their editing skills, practices, and reflections: Jon Junia, Angela Vanduinen, Casey Rudden, Lindsey Latour, and Robin Pulford. Your insight gives life to this chapter.

REFERENCES

Albers, M. J. (2000). The technical editor and document databases: What the future may hold. *Technical Communication Quarterly, 9*(2), 191-206.

Allen, L., & Voss, D. (1998). Ethics for editors: An analytical decision-making process. *IEEE Transactions on Professional Communication, 41*(1), 58-65.

Anderson, S. L., Campbell, C. P., Hindle, N., Price, J., & Scasny, R. (1998). Editing a web site: Extending the levels of edit. *IEEE Transactions on Professional Communication, 41*(1), 47-57.

Business-Higher Education Forum. (1999). *Spanning the chasm: A blueprint for action.* Washington, DC: American Council on Education and National Alliance of Business.

Caruthers, C. (1986). The evolution of a style guide. In *Proceedings of the 33rd International Technical Communication Conference* (pp. 405-407). Washington, DC: Society for Technical Communication.

College Board. (2004). *Writing: A ticket to work . . . Or a ticket out. A survey of business leaders.* Available:
http://www.writingcommission.org/prod_downloads/writingcom/writing-ticket-to-work.pdf

Dayton, D. (2003). Electronic editing in technical communication: A survey of practices and attitudes. *Technical Communication, 50*(2), 192-205.

Dayton, D. (2004a). Electronic editing in technical communication: The compelling logics of local contexts. *Technical Communication, 51*(1), 86-101.

Dayton, D. (2004b). Electronic editing in technical communication: A model of user-centered technology adoption. *Technical Communication, 51*(2), 207-223.

Dragga, S., & Gong, G. (1989). *Editing: The design of rhetoric.* Amityville, NY: Baywood.

Farkas, D. K., & Poltrock, S. E. (1995). Online editing, mark-up models, and the workplace lives of editors and writers. *IEEE Transactions on Professional Communication, 38*(2), 110-117.

Gaidosch, B. (2003, August 9). Career climbers need better writing skills (AP). *The San Diego Union-Tribune.* Distributed by KnightRidder/Tribune Business News.

Gerich, C. (1994). How technical editors enrich the revision process. *Technical Communication, 41*(1), 59-70.

Grensing, L. (1988). Building your business writing. *Management World, 17*(2), 17-18.

Gross, G. (Ed.). (1993). *Editors on editing: What writers need to know about what editors do.* New York: Grove Press.

Haugen, D. (1991). Editors, rules, and revision research. *Technical Communication, 38*(1), 57-64.

Leininger, C., & Yuan, R. (1998). Aligning international editing efforts with global business strategies. *IEEE Transactions on Professional Communication, 41*(1), 16-23.

Nichols, M. C. (1994). Using style guidelines to create consistent online information. *Technical Communication, 41*(3), 432-438.

Plotnik, A. (1982). *The elements of editing.* New York: Simon & Schuster Macmillan.

Read, M. (2004). Employers urge workers to improve writing (AP). Available: http://www.ledger-enquirer.com/mld/ledgerenquirer/news/nation/9659764.htm

Rubin, C. (n.d.). "You are what you write," Guerilla Marketing Articles. Article on the Guerilla Marketing.com web site. Accessed April 25, 2005 at http://www.gmarketing.com/articles/read/103/You_Are_What_You_Write.html.

Rude, C. (2002). *Technical editing* (3rd ed.). New York: Longman.

Samson, D. C., Jr. (1993). *Editing technical writing.* New York: Oxford University Press.

U.S. Department of Labor. (1999). *Futurework: Trends and challenges for work.* Available: http://www.dol.gov/asp/programs/history/herman/reports/futurework/report.htm

Williams, T. R., & Harkus, D. A. (1998). Editing visual media. *IEEE Transactions on Professional Communication, 41*(1), 33-46.

Windhaus, S. (2004, August 10). *Our age is showing?* Forum post on Small Business Computing.com. Retrieved April 25, 2005 at http://forums.smallbusinesscomputing.com/showthread.php?t=393.

RECOMMENDED READING

Allen, P. R. (1996). User attitudes toward corporate style guides: A survey. *Technical Communication, 43*(3), 237-243.

American Psychological Association. (2001). *Publication manual of the American Psychological Association.* Washington, DC: American Psychological Association.

Anderson, L. K. (1990). *Handbook for proofreading.* Chicago: NTC Business Books.

Bocchi, J. (1998). Technical editing in transition: Editors wanted for intranet site development. *IEEE Transactions on Professional Communication, 41*(1), 5-15.

Brooks, B. S., & Sissors, J. Z. (2001). *The art of editing* (7th ed.). Boston: Allyn & Bacon.

Bush, D. W., & Campbell, C. P. (1995). *How to edit technical documents.* Phoenix, AZ: Oryx Press.

Caernarven-Smith, P. (1991). Aren't you glad you have a style guide? Don't you wish everybody did? *Technical Communication, 38*(1), 140-142.

Chicago Manual of Style (15th ed.). (2003). Chicago: University of Chicago Press.

Eisenberg, A. (1992). *Guide to technical editing: Discussion, dictionary, and exercises.* New York: Oxford University Press.

Einsohn, A. (2000). *The copyeditor's handbook: A guide for book publishing and corporate communications.* Berkeley: University of California Press.

Mackay, P. D. (1997). Establishing a corporate style guide: A bibliographic essay. *Technical Communication, 44*(3), 244-251.

Nadziejka, D. E. (1995). Needed: A revision of the lowest level of editing. *Technical Communication, 42*(2), 278-283.

Perkins, J., & Maloney C. (1998). Today's style guide: Trusted tool with added potential. *IEEE Transactions on Professional Communication, 41*(1), 24-32.

Skillin, M. E., & Gay, R. M. (1974). *Words into type.* New York: Prentice Hall.

Washington, D. A. (1993). Creating the corporate style guide: Process and product. *Technical Communication, 40*(3), 505-510.

CHAPTER 17

Exploding the Myth of Transparent Communication of Information

Marilyn M. Cooper

> **OUTCOME:** Students will understand that clear communication of information depends on evoking the contexts within which documents are used and on working with readers to understand their needs.

INTRODUCTION

The communication of information is often viewed as a simple process involving accuracy, objectivity, completeness, and most of all, clarity of language, sentence structure, organization, and images. In Houp, Pearsall, Tebeaux, and Dragga's popular textbook, clarity tops the list of skills important for communication in the workplace (2002, p. 4) and is listed as the "most important attribute" of good technical writers: "Until the sense of a piece of writing is made indisputably clear, until the intended reader can clearly understand it, nothing else can profitably be done with it" (p. 6). But this ideal of clear communication rests on a misunderstanding of language as transparently representing the real world—what Carolyn Miller referred to as the "windowpane theory of language" (1979, p. 610)—according to which the world can either be represented as it clearly is or in a way distorted by the beliefs and intentions (or carelessness) of the communicator. As our society and the global world of work has become insistently more heterogeneous, inhabited by people with widely divergent cultures and values, the naïveté of the windowpane theory has become obvious.

Instead, communication needs to be seen as a stimulus, a difference that provokes responses in others, inviting them to become involved in the construction of useful understandings. Informing consists of offering a perspective different from that already held by the readers and thus stimulating them to respond to that difference. As responses vary, understandings will vary and will not be predictable. This, however, is the value of informing, for if readers already understand the phenomena, events, and technologies as the writer does, there is no information to be communicated. The purpose of clarity in the communication of information is to enable readers to interact with the world safely, effectively, and productively.

WHAT CORPORATIONS AND GOVERNMENT AGENCIES TELL US

Clarity in communication is a prime requirement for employees of corporations and government agencies. The Department of the Interior, in its Benchmark Employee Performance Standards, defines exceptional employees as producing "exceptionally clear and effective" written and oral communications and defines minimally successful employees as demonstrating "lack of clarity in writing and speaking." In a survey of business leaders conducted in 2004 by the College Board's National Commission on Writing, respondents overwhelmingly chose accuracy (95.2%) and clarity (74.6%) as "extremely important" characteristics of written communication.[1] One respondent commented, "Manufacturing documentation, operating procedures, reporting problems, lab safety, waste-disposal operations—all have to be crystal clear" (2004, p. 8).

Further testifying to the importance placed on the clarity of public documents was the emergence of the plain English movement. In 1975 the Magnuson-Moss Warranty Act required all warrantees to be written in plain language, and in 1998 the Securities and Exchange Commission adopted a rule that required prospectuses to be written in plain English and published *A Plain English Handbook: How To Create Clear SEC Disclosure Documents.* In the preface, Warren E. Buffett, who supported the work of creating the handbook, comments that when he has studied documents that public companies file, "Too often, I've been unable to decipher just what is being said, or worse yet, had to conclude that nothing was being said" (1998, p. 1).

Clear communication saves time and money and improves comprehension. Yvonne Alexander explains how unclear writing can lead to financial losses for a company by instancing an HMO whose suit to collect a repayment from a client was denied by a judge "based on the fact that the policy was written so badly that no one could understand it" (qtd. in Tyler, 2003). Joseph Kimble (1996-97) summarizes 25 studies and reports that demonstrate the cost-effectiveness of revising documents into plain English. For example, when the FCC revised its regulations for marine radios on recreational boats:

---

[1] The Commission sent surveys to "human resource leaders" of 120 "major American corporations" and received 64 responses. The sample was not representative of the whole of the workforce, as no responses were solicited from government, small businesses, and the wholesale/retail trade; manufacturing was greatly overrepresented (69.2% of respondents for an area that employs 16.2% of American workers); and services was underrepresented (6.4% of respondents for an area that employs 28.8% of American workers).

> Readers of the old rules got an average of 10.66 questions right out of 20; readers of the new rules got an average of 16.85 right. The average response time improved from 2.97 minutes to 1.62 minutes. Finally, on a scale of 1 (very easy) to 5 (very hard), readers rated the old rules at 4.59 and the new rules at 1.88 (p. 26).

Kimble notes that the significant improvement in these rules resulted from adhering to "what may be the hardest principle of all to follow, because it involves judgment and restraint—don't try to cover every remote possibility under the sun" (p. 26).

Clarity is a concern in composing all aspects of a document, including organization, formatting, and visuals. Comments on clarity often focus on sentence structure and word choice, but efforts to improve the clarity of workplace writing usually include all aspects of communication. For example, here's how a book about e-learning defines clear communication:

> Clear writing means designing materials that are easy to use because they are organized logically. It means using concrete, familiar language that sounds like every-day speech. It means formatting documents to make information accessible, easy to read, to follow and to visualize. Finally, clear writing means choosing illustrations that are clear, memorable and sensitive to readers' concerns and needs (e-LearningHub.com, 2002).

Echoing this definition, Kimble emphasizes that

> plain language is not just about vocabulary. It involves all the techniques for clear communication—planning the document, designing it, organizing it, writing clear sentences, using plain words, and testing the document whenever possible on typical readers (1996-97, p. 2).

And the plain English rule adopted by the SEC stipulates that among other things, prospectuses contain clear, concise sections, paragraphs, and sentences; bullet lists, headings; and clear visuals.

Clarity depends fundamentally on understanding readers' needs. The definitions quoted above notably draw attention to the needs of readers. Though the bulk of the *Plain English Handbook* (38 of 64 pages) is devoted to tips culled from standard editing, style, and design guides, the handbook agrees that achieving the goal of clearly presenting complex information means first understanding readers' needs. In his introduction, Arthur Levitt, Chairman of the SEC, comments, "We must question whether the documents we are used to writing highlight the important information investors need to make informed decisions" (1998, p. 3).

It is particularly difficult for writers to judge the appropriate level of detail needed by readers, to include the details readers need but avoid covering "every remote possibility under the sun." When the physicists at the Fermi National Accelerator Laboratory committed to posting on their web site a plain English version of all their technical papers, readers of early attempts complained that the writers "assumed the reader already knew what they were talking about." A teacher who uses science writing in her composition classes explained that the writers didn't understand what their readers needed to know and expected too much of them: "If you can't gauge your audience, you have no way to figure out what level of detail you need to go into when you define your terms" (Glanz, 1997).

The best way to gauge what readers need is through testing documents with readers. Usability testing, because it involves potential readers in the writing process, results in documents that are much more likely to address readers' needs. Discussing a project that used plain English principles to revise 100 form letters for the Washington State Department of Labor and Industry, Dana Howard Botka (2002) reports that asking typical readers to comment on drafts of form letters resulted in substantial revision and improvement. Kimble (1996-97) argues for using a variety of methods to ascertain that readers' needs are being satisfied. He notes that the team that revised Federal Express ground-operations manuals "took all the steps: they did a field study of users, tested the old manuals for usability, and compared the manuals to benchmark standards" (p. 14). He concludes, "The cardinal rule of clarity is to put yourself solidly in the minds of your readers: what would they like to know, and how would they like to get it?" (p. 26).

Formal methods to ensure clarity, such as readability formulas or computerized style checkers, are a poor way to gauge clarity because they cannot address content or context. The penultimate chapter of the *Plain English Handbook* (1998) warns against reliance on such methods, cautioning that readability formulas cannot determine whether information has been conveyed clearly and that the suggestions of computerized style checkers must be taken only as suggestions. Reporting on their research with a medical technology corporation, **Giles and Still note that technical communicators in industry agree that readability is best determined by usability studies and not by readability formulas (2005, pp. 50-51).**

WHAT ACADEMIC RESEARCH TELLS US

The goal of achieving absolute clarity through textual features rests on a misunderstanding of language as transparently representing the real world. Earl Britton's influential definition of technical writing, quoted in Miller (1979), is based on the windowpane or transmission model of language and emphasizes lack of ambiguity as central:

> The primary . . . characteristic of technical and scientific writing lies in the effort
> of the author to convey one meaning and only one meaning in what he says.
> That one meaning must be sharp, clear, and precise. And the reader must be
> given no choice of meanings (1979, p. 11).

Miller argues that this model is based on the positivist view of science and has dominated the definition and teaching of technical writing for too long. Just as most thoughtful scientists have rejected positivism, she says, so too should technical communicators reject the positivist theory of language and instead understand all writing, including technical writing, more rhetorically, as an activity in which humans construct versions of reality and attempt to persuade others that these versions are useful in understanding the world. On this understanding, she says, "Good technical writing becomes, rather than the revelation of an absolute reality, a persuasive version of experience" (1997, p. 616).

Jack Bushnell (1999), who returned to teaching from working in industry, echoes her argument in his report of his experience:

> I learned in those professional settings that there is no such thing as a neutral, merely informative document. I learned that all writing is "positioning," that an ostensibly descriptive report has persuasive intentions, that an instructional manual (including software documentation) reveals its cultural and political assumptions repeatedly (p. 176).

He notes that the assumptions of most textbooks and even most scholarship in technical communication conflicts with his experience, pointing out that the 9th edition of Houp, Pearsall, Tebeaux, and Dragga (2002) still lists objectivity and clarity as two of the five attributes of good technical writers and arguing that "in doing so, it echoes the same positivist faith in neutrality and the transparent nature of language that characterized the very first edition of the book thirty years ago" (p. 177; these attributes remained the same in the 10th edition, published in 2002).

Clarity rests instead on evoking contexts that enable useful interpretations. Others, including myself (see Cooper, 1996; Dobrin, 1989; Mirel, 1998; Weiss, 1993), have argued that because communication constructs reality rather than revealing it, clarity cannot be achieved merely through choosing language and images. In understanding texts, audiences bring to bear a wealth of background information, common sense, assumptions, values, and expertise to interpret the signs on the page or screen. Technical communicators work to evoke aspects of these contextual knowledges and abilities that will enable audiences to integrate new information and perspectives into their thinking and action. Clarity thus relies on cooperative interactions between writers and readers in which mutually beneficial meanings are negotiated.

David Dobrin explains that clear directions and instruction rely on enabling readers to organize their Background skills and network of knowledge in an appro-priate way. While the relevant network can be cued directly by including details of knowledge (although which details to include can be tricky to ascertain), Background skills—which include things like noticing unusual features, not immersing electrical appliances in water—"cannot be imparted by relating facts" but must be evoked by "conveying information that can be used by people with the right Background skills" (1989, p. 52). In instruction manuals, he argues, Background skills can be evoked by getting the user "to adopt the point of view of the designer" (p. 26), which largely depends on explaining how the technology is intended to be used. Van Horen, Jansen, Maes, and Noordman's (2001) report of a study of difficulties elderly people have in using manuals supports Dobrin's analysis, concluding that the absence of information about goals, along with an absence of information about the consequences of actions and identifications or descriptions of the parts referred to, has a negative effect on task performance.

Barbara Mirel (1998), also addressing the problems of effective user documentation, similarly explains that writers must evoke tacit knowledge that users need to perform new, complex activities. Tacit knowledge cannot be usefully transformed into "explicable rules, propositions, steps, and strategies . . . for learners to emulate" because doing so abstracts this knowledge from the "lived experiences of work" (p. 30). Thus, she suggests problem-based instruction using cases which "leave space (narrative silence) for users to fill in with their own contextually dependent interpretations and choices" (p. 44). (For more on using cases, see Chapter 6.)

For communicating clearly in a global economy and in multicultural societies like our own, it is even more essential to leave spaces open for users to fill with their own contextually dependent interpretations. Timothy Weiss traces the fixation on clarity in American technical communication to the post-World War II intellectual paradigm in which the

> climate of burgeoning bureaucracy and black-and-white ideologies surely influenced the choice of principles that business and technical communication enshrined, such as directness and clarity in the face of bureaucratese. These principles also reflected the United States' economic and cultural dominance, isolationism, and preference for simplicity (1997, p. 325).

In many other cultures, clarity and directness are instead considered to be childish and therefore insulting to readers. Weiss comments, "We need not talk to one another as if we were children . . .—yes, children need clarity—but adults can deal with indirectness and multiple meanings" (1997, p. 324). He argues that the central problem in technical communication "is often not one of clarity (simply focusing the camera right, so to speak) but of understanding (interpreting) and formulating the understanding (translating) into appropriate messages" (1997, p. 330). Arguing that the transmission model "cannot make sense of communication in a global society of distinct cultures and their distinctly different interpretations of the world" (1993, p. 205), he argues that "difference is the very stuff of communication, or more precisely, that communication depends on a sense of commonality *and* difference" (1993, p. 204). He contends that a message is not transmitted between participants, but, as Medvedev and Bakhtin say, "is constructed between them as a kind of ideological bridge, is built in the process of their interaction" (1978, p. 152; see Weiss, 1993, p. 206).

As these analyses suggest, **clear communication is more likely when readers participate dialogically in the creation of meaning,** something that Barbara Schneider suggests, too. She does not advocate abandoning traditional advice, but she argues that

> a writer can consider her audience, write grammatically, achieve technical accuracy, and still produce something that is unclear to the reader. We must

therefore also be aware that the nature of language makes misunderstanding an inevitable aspect of communication and that shared context must be constantly and explicitly created and maintained (2002, p. 213).

Misunderstandings should be expected even between people who have worked together in the same workplace for a long time. Schneider goes on to offer an example of a conversation at her university between two administrators of a mandatory writing test for students. The director asked if they were still going to post the results of the test; the assistant answered that they were required to post the results; and the director responded, "I meant to post them on a bulletin board" (p. 211). The salient context for the director was their move to a new, smaller office, while the salient context for the assistant was her work of putting the results of the test into the computer. Because misunderstandings like this are extremely common, repair work is a regular part of everyday conversation.

The challenge in technical communication documents is finding ways to do such repair work in advance. Again, Schneider offers a number of recommendations, such as using introductions and cues that establish and maintain the context in which the document is to be used, providing contact information for readers with questions, meeting and talking with the recipients of reports and memos to explain the contexts and contents of the documents, and involving readers in the writing of documents. This last is a method of usability testing, and there is no better way to assure the clarity of a written document than testing it with actual readers. (For more on usability testing, see Chapter 11.)

A note on readability formulas. Giles and Still comment that though technical communicators consider readability formulas to be unreliable, they "maintain a certain ubiquity," included in word-processing programs and frequently required by contracts with government agencies and corporations (2005, p. 47). In their work with a medical technology corporation, one of them asked the technical writers at the corporation "why they didn't simply explain to the government that most formulas do not render a meaningful evaluation. They found his naïveté amusing" (p. 68). Technical communication students, thus, should be familiar with standard readability formulas but clearly understand their limitations. Because such formulas rely only on the number of syllables per word and the number of words per sentence, they militate against strategies of clear communication such as sentence combining. Giles and Still tested the Flesch-Kincaid formula on a revised manual for a pacemaker. One sentence was revised from

In general, keep AC-powered, hand-held devices several inches away from your pacemaker. This will reduce the possibility of EMI.

to

In general, keep AC-powered, hand-held devices several inches away from your pacemaker to reduce the possibility of EMI.

Although the second version eliminates the ambiguous "this" that begins the second sentence, the Flesch-Kincaid readability formula rated the first version as ninth-grade level and the second as tenth-grade level. Giles and Still recommend that when readability formulas are required, technical writers use a formula that measures aspects of sentence structure such as the Golub Syntactic Density Score; this formula rated both sentences as third- to fourth-grade level reading.

ASSIGNMENT SEQUENCE

These exercises and assignments encourage students to achieve clarity through understanding and responding to readers' needs and by evoking shared contexts and involving readers in the writing process. The first in-class exercise problematizes students' commonsense understanding of clarity. Two other in-class exercises demonstrate the context-sensitivity of strategies that aim to improve clarity through restructuring sentences or reorganizing information. The project assignment requires students to develop and practice methods of evoking and maintaining shared contexts and to engage in various methods of usability testing to produce a document that enables readers to use information effectively and safely.

Sensitivity Exercise

This in-class exercise uses experiential learning to undermine the commonsense approach to achieving clarity through unambiguous language. Briefly, the activity asks students to describe a person so accurately that he or she can be identified solely on the basis of the description. To make the exercise more interactive, it is set up as a contest.

Set Up

Explain to the class that the activity for the day is designed to test their abilities to communicate information accurately, and that to increase their incentive to do well, they will be competing in two groups. Choose one student from each group to be the "target" of description. Do not choose a student who obviously stands out—the only white male, the only female wearing a red sweater, for example.

Writing Assignment

Ask each group to write a description of their target member that will enable the members of the other group to identify him or her on the first try. The target should be involved in the writing process. Separate the two groups so they cannot overhear each other's discussions.

(Con)Test

Reconvene the class. Tell the students that each group will take turns in identifying the target and that the description that enables the first correct identification wins. If both groups take the same number of tries to identify the target, the contest will be declared a tie. Give one description to one group, asking them to read and discuss it out loud and then to make a choice. Repeat with the other group. Keep alternating groups until both have correctly identified the target. Award prizes (coffee coupons, candy).

Discussion

To elucidate why clear and accurate communication can be difficult, ask students to discuss these questions:

- How did you decide what information to include in your description?
- Why did you think the information you chose would enable the target to be quickly identifiable?
- What choices in words, sentence structure, and organization did you make to make your description more successful?
- What information in the description you read was confusing or ambiguous?
- What background knowledge did the description you read seem to rely on?
- What was confusing or misleading in the words, sentence structure, or organization of the discussion you read?
- What added information would have enabled you to identify the target more quickly?

The important part of the exercise is the type of insights that come out of the groups' discussions as they write and read the descriptions and that come up in the reflective discussion. Emphasize in this discussion how difficult it is to figure out how others will interpret the texts, especially what contextual or background information they will rely on. If your class has some diversity, you can highlight differences in cultural expectations too.

Exercise in Clear Writing

This exercise focuses on the necessity of understanding readers' needs in order to effectively use strategies for clear writing offered by handbooks and style guides. It takes two class sessions.

Class Session 1

Assemble lots of two-to-three-sentence passages of writing that lack clarity because of their structure. Look for diverse examples of each of the strategies for clear writing you want to cover. In class, present passages to the class and ask them how the passages might be revised to make them clearer. Ask them why they make particular revisions to enable them to see patterns in their suggestions and articulate strategies they unconsciously use or have been explicitly taught. When students suggest different revisions, ask them what the different revisions assume about the readers or about the context of the communication. At the end of class, give students a reading on clear writing (a handout I use in my writing and editing classes is found in Appendix A), telling them to add notes on any additional strategies that came up in discussion. For the next class session, give them a passage to revise for clarity. (Appendix B contains a passage for revision.)

Class Session 2

Go over the passage sentence-by-sentence, asking students how and why they revised as they did. Focus on what assumptions about the readers and the context of the passage underlie the suggested revisions, and postulate other kinds of readers and contexts that would necessitate different revisions. For example, most students will revise the first sentence of the passage to avoid the extensive nominalization (and also to get the framework at the beginning) to a structure like this:

> **If executing an operation will not disable any other operation that needs to be executed, the procedure is sequentially unconstrained.**

Or, more informally,

> **If you do one operation that needs to be done and that does not keep you from doing another operation that needs to be done the procedure is sequentially unconstrained.**

The unrevised version of this sentence, though it does contradict some of the strategies suggested for clear writing, appears as it does in the original book (as does the parallel first sentence of the second paragraph). The intended readers of the book are cognitive anthropologists, and the nominalization allows the writer to define the terms "sequentially unconstrained" and "sequentially constrained" with precision and economy. Putting the framework at the end is also a choice that signals the intention to define terms. In revising, students assume a reader more like themselves, perhaps one who wants to learn how to efficiently perform various procedures. The change to more verbal forms ("executing" rather than "the execution of" and "that needs to be executed" instead of "enabled") reflects an assumption that the reader is interested in doing the procedures rather than in defining differences among types of procedures. The second revision strengthens the assumption of a lay reader by inserting "you" as the agent and also changing formal terms like "executing" and "disable."

Another interesting point to discuss is which readers will find the two examples—orthodox dressing and the interlock on automobile starting engines—effective and which readers will not.

To end the class with an interactive demonstration of the same lesson, ask students to rewrite the following sentence in the fewest number of words possible.

> Because of the fact that flowers have occasion to come into bloom at this point in time, it is necessary for us to give due consideration to the possibility that

class sessions could be convened in closer proximity to the natural environment.
(42 words)

Have them compare their revisions, and again focusing on assumptions about the
readers and context, ask students to discuss why they thought certain information
could be left out.

Critique of Instructions

This exercise focuses on how the explicitness and level of detail needed in
instructional writing depends on the intended audience. It is a single-class-session
exercise that engages students in group discovery of how differences in the
knowledge and experience of a reader affects the clarity of written instructions.
It also demonstrates how difficult it is to write instructions that are clear to a wide
range of readers.

Preparation

Ask students to find, analyze, and bring to class an instruction manual that they
think is confusing. They should consider in their analysis what information needs to
be added or explained or what reorganization is needed to make the manual clearer
as well as considering what features of the manual are helpful.

Group Discussion

To maximize differences in levels of knowledge among group members, divide
the class into groups in such a way that each group considers manuals for a variety
of technologies. Ask each student to show their manual to their group members
and explain their analysis. After each analysis, group members should discuss the
extent to which their reading of the manual agrees with the presenter, considering
these questions:

- Did they find the manual confusing?
- Did they find the same things in the manual confusing? Or other things?
- Did they need the information the presenter suggested adding? Or other
  information?
- Did they find any suggested reorganization useful?
- Did they find the manual confusing because it had too much detail?
- Did they find the same features helpful? Or other features?

After all the manuals have been discussed, groups should discuss what caused the
differences and similarities in how they read the manuals and should list the kinds
of things writers of instructional manuals need to know about their intended readers
in order to produce a useful document.

As you check in on the groups while they work, encourage them to focus on the
differences in group members' prior knowledge of the technologies being discussed

and on the level of knowledge assumed by the manuals. If a group is having difficulty, draw their attention to the technical terms being used and to steps that assume what Dobrin (1989) calls Background skills. For example, camera instructions often assume that readers know what the terms "aperture" and "shutter speed" mean and how they affect the image, and Backgound skills assumed by many electronic devices include paying attention to and discriminating between icons, as well as selecting text or labels before performing an operation.

Whole Class Discussion

Ask each group to discuss their observations about what writers of instructional manuals need to know about their readers. Then ask the class to discuss what strategies—such as arranging information differently or enabling readers to navigate the manual differently—writers can use to address differences in readers' need for information and level of detail. Draw students' attention to the use of visuals, analogies, examples, cultural references, and metaphors as ways of evoking Background skills and the correct contexts for interpretation. If suggestions like Dobrin's (getting the reader to adopt the point of view of the designer) or Van Horen et al.'s (2001) (include information about goals, consequences of actions, and iden-tification of parts) do not come up, explain them to the class, using examples from the manuals the students have brought in.

Project Assignment

This is a major assignment that asks students to focus on ways to involve readers in constructing and testing texts as means of ensuring clear communication. It can easily be combined with the assignment sequence offered in Chapter 6 or the first project assignment in Chapter 2. It requires students to write for real, not hypothetical, readers who will be using the document created by student authors and with whom students can work directly. It also requires students to write a document for an organization that they participate in, so that they will have contextual knowledge of the organization and be able to recruit readers to help them with their work. And like many of the assignments in this book, it requires students to reflect frequently on their work in process and on what they are learning from doing this work.

The assignment sheet is reproduced in Appendix C. Besides assuming students have discussed topics mentioned above and have done the three in-class exercises, the assignment assumes that students have some understanding of how to work rhetorically with the design of texts and visuals (see Chapter 4).

As students work through the stages of the assignment, you will need to spend class time discussing their understanding of the particular skills required, such as interviewing, literature research, and usability-test design. It's most effective to work inductively in these discussions, asking students what they've been taught about these skills, what their experiences have been, and what strategies have worked for them. Even in introductory classes, students have a lot of knowledge of these skills, and finding out what they already know—and what misconceptions they have—can save time. For further information on the usability tests required in stages 4 and 5 of the assignment, see Chapter 11.

Assessment

The rationale for basing the grade for the project on students' reflections and final commentary is that the purpose of the assignment is to help students learn how to create clear, usable documents, not to test whether they can already do so. The creation of one good document is not a particularly good measure of students' ability. Students who can articulate what they have learned and what strategies go into producing usable documents are more likely to be able to consistently produce them.

Getting student input on the evaluation of projects is also a valuable learning experience. Because in this project students are creating documents that will vary greatly in purpose, genre, and format, devising a rubric for assessment with the class (as described in Chapter 4) is not feasible. I have used grading conferences in which I ask students to discuss the extent to which they have achieved the goals of the assignment and to suggest a grade for their work. I then respond to their assessment of their work and tell what grade I would assign. If their suggested grade is higher— or as is more often the case, lower—than the one I suggest, we negotiate.

CONCLUSION

Working with sentence structure and word choice to improve clarity can be concrete and thus satisfying (at least for those who like working with words). But poorly structured sentences rarely evoke in readers anything more than annoyance. Misunderstandings and incomprehension of the kind that leads to documents being unusable results primarily from global disorganization, the provision of too little—or too much—information, and the failure to evoke in readers the context within which the document is to be used and in which it makes sense. Clear communication depends on writers paying attention to these matters, and doing so necessitates paying attention to the specific needs of the real readers of the document. Technical writers have developed many strategies for working with the real readers of documents, and I have touched on only a few here. As you and your students work on these assignments, you probably will discover new strategies. You will also be able to find more strategies in the professional literature on technical writing. Much of the literature on working with readers overlaps the discussion of usability testing, and indeed, clear communication is more accurately thought of as a matter of ensuring the usability of documents rather than of achieving transparency in language.

APPENDIX A
Clear Writing

Careful arrangement of information in a sentence and across sentences can help readers understand the intended message and emphasis more clearly, but using these suggestions successfully depends on your thorough understanding of what readers know and believe and what they need to know and believe. And note that these are suggestions, not rules to be followed slavishly.

old information; new information: One general pattern to follow is to put "old" information first in the sentence and new information second. Old information is information that has been introduced earlier in the text or information that is well known. New information is what news is being added.

In the following sentences, old information is **bold** and new information is underlined.

> **The training lead**, in my opinion, is the most important piece of equipment that I ever use, but, like any **training aid, it** has to be used sensitively. **The lead** should be used like a fishing line—and as gently as a fishing line.

framework at the beginning: Putting adverbs or prepositional phrases that state when, where, or how the event referred to in the sentence took place at the beginning of the sentence allows readers to put the information in context.

> **Since the early days of biology**, philosophers and scientists have noticed that living forms have stable structures yet change continually.

> **In the process of photosynthesis**, solar energy is converted into chemical energy and bound in organic substances.

> **Surprisingly**, most of the substance of green plants comes from the air.

linking words at the beginning: Similarly, putting linking words (conjunctive adverbs) at the beginning allows readers to understand immediately how the information in the sentence relates to the previous text.

> In a binary network of 100,000 genes, the possibilities of different patterns of gene expression are astronomical. **However,** the number of attractors in such a network at the edge of chaos is approximately equal to the square root of the number of its elements. **Therefore,** a network of 100,000 genes should express itself in about 317 different cell types.

emphasis at the end: Readers generally expect to get context first, followed by what is new, so information you want to emphasize should be placed at the end of the sentence. Notice what happens with the rearrangement of information in the following passage.

The lead should be left trailing on the ground in later training stages. You can just step on the lead to prevent the repetition of mistakes when the lead is on the ground.

In later training stages, the lead is left trailing on the ground. **With the lead on the ground, preventing the repetition of mistakes** is simply a matter of stepping on the lead.

lists at the end: If a sentence presents a list, it's easier for readers if they know the context of the list first.

> **Not:** True shapes of physical features, correct angular relationships among positions, equal area, or the representation of areas in their correct proportions, and constant scale values are the desirable properties of charts used in navigation.
>
> **Better:** The desirable properties of charts used in navigation are: true shapes of physical features, correct angular relationships among positions, equal area, or the representation of areas in their correct proportions, and constant scale values.

longest item at the end of a list: If one item is longer or more complex, it's easier for readers if the long item is at the end of the list.

> **Even better:** The desirable properties of charts used in navigation are: true shapes of physical features, correct angular relationships among positions, constant scale values, and equal area, or the representation of areas in their correct proportions.

action verbs: Naming the action in the verb usually makes a text easier to understand than naming the action in a noun.

> **Not:** The basic nutrient elements **make an appearance** in a variety of chemical compounds.
>
> **Better:** The basic nutrient elements **appear** in a variety of chemical compounds.

doublets: Using two words that are similar in meaning joined by **and** or **or** is often less clear than using the one most appropriate word.

> **Not:** Tasks that have sequential constraints require some **planning or coordination** among the actions to be taken.
>
> **Better:** Tasks that have sequential constraints require some **coordination** among the actions to be taken.

> **Not:** Many of the elements of the preparation for Sea and Anchor Detail **involve and require** the performance of the parts of the anticipation computation.
>
> **Better:** Many of the elements of the preparation for Sea and Anchor Detail **involve** the performance of the parts of the anticipation computation.

phrases to avoid: These phrases add words but add nothing to the meaning. Delete them or replace them with a word or shorter phrase.

- the fact that
- as far as
- as long as (since)
- at this point in time (now)
- in this day and age (today)
- it is the belief of many workers that (many workers believe)
- there are many who like the change (many like the change)

APPENDIX B
Revise for Clarity

Using the strategies discussed in class, revise the following passage to make it as clear as possible. Be sure to think carefully about who the readers of this passage are likely to be and what they know and need to know.

Sequential Control of Action

A procedure is *sequentially unconstrained* if the execution of any enabled operation will never disable any other enabled but as yet unexecuted operation. A "swarm of ants" strategy or tactic is the way to accomplish a task that has no sequential constraints. There is no communication between the active agents other than their effects on a shared environment in such a scheme. Each agent simply mills about taking actions only when situations on which he can act are encountered.

A procedure is *sequentially constrained* if the execution of any enabled operation will disable any other enabled but as yet unexecuted operation. It is a fact that it is necessary to have some control over the sequence of actions where there are sequential constraints.

The performance of a sequentially constrained procedure may require planning or backtracking. Getting dressed is sequentially constrained for example, because at the moment in which one has neither shoes nor socks on, putting on shoes disables the operation of putting on socks. The sequence of operations for orthodox dressing contains a sequential constraint on the donning of socks and shoes.

The manipulation of the enablement conditions of various operations is one general technique for turning sequentially constrained tasks into sequentially unconstrained tasks. A simple rule is "suppress the enablement of any operation that could disable another already enabled operation." This can be done through interlocks. The starter motor will not turn unless the transmission is in park or neutral in many automobiles, for example. This is the mechanical enforcement of a sequential constraint on the engine-starting procedure.

(Adapted from Edwin Hutchins, *Cognition in the Wild,* Cambridge: MIT Press, 1995, 198-199.)

APPENDIX C
Project Assignment: Writing for Readers

The goal of this project assignment is to enable you to discover and practice strategies for working with readers to create documents that fulfill their needs. It requires you to create a document for an organization of which you are an employee or member—a business enterprise, club, church, or volunteer group, for example. The document should contain at least 250 words and should include visual elements. The genre of the document and its format should fit the purpose of the document.

In creating the document, you will work through a number of stages that involve interviewing, focus groups, and reader testing as well as drafting and redrafting. As you finish each stage, in addition to reports and drafts of the document, you will write a reflection on what worked, what didn't work, and what you learned about writing for your target readers. The purpose of these reflections is to enable you to articulate and remember useful practices and strategies for writing documents that fulfill readers' needs. The last stage asks you to write an extended reflection—a critical commentary on all your work on the project and a discussion of principles for effectively attending to readers' needs in documents. **Your grade for the project will be based largely on the learning you demonstrate in your reflections and final commentary.**

Stage 1: Proposal

The first step is to propose what kind of document you will create for what organization. Choose an organization that you participate in as an employee, volunteer, or member and that you are fairly familiar with. Think about what documents this organization needs right now, what the purpose and topic of the document should be, and what genre (informational piece, instructions, narrative, argument) and format (brochure, blog, web site, report, memo) would best suit these needs. Talk with other organization participants and administrators or officers as you decide what document to create.

Write a **proposal** that includes the following information:

- The purpose and topic of the document to be created
- The genre and format of the document
- The organization for which the document is to be created
- How and how long you have participated in this organization
- Why this is a valuable document to create for this organization

Write a **reflection** in which you discuss how you decided on what kind of document to create. How did your understanding of the institution help you make the decision? Why did you reject certain ideas? What did you learn from other organization participants and administrators or officers that influenced your decision?

Stage 2: Report of Research

Next you need to do some research on what information and impression the organization wants the document to communicate, how readers will use the document, and what the requirements and standards are for the genre and format you will use in creating the document. You will need to

- Interview organization administrators or officers: What do they see as the purpose of the document? What information needs to be included? What impression of the organization do they want the document to convey?
- Interview potential readers: What do they see as the purpose of the document? What information should the document contain that would enable them to use it effectively for this purpose? What features would they find useful in the document?
- Do research to develop the content of the document: Collect and read other documents used in the organization; research the topic of the document in technical reports, periodicals, web sites; interview experts on the topic.
- Do research on the requirements and standards for the genre and format you will use: Assemble examples of the genre and format, and reflect on what features are required and useful; give your examples to classmates and ask them what features they find useful; research guidelines for the genre and format in textbooks, professional handbooks, and style manuals; research standards for the genre and format in professional and technical writing periodicals.

Write a **report** in which you summarize the results of your research and discuss how you will use it in the creation of your document. Include citations for all sources, including the people you interviewed, and for all documents and examples you collected.

Write a **reflection** in which you discuss what worked well for you in your research process, what problems you encountered, and what strategies you would use in the future to improve the effectiveness and efficiency of your research. Be sure to discuss how you prepared for and conducted interviews and what search strategies you employed in finding sources, documents, and examples.

Stage 3: Testing First Draft

Create a first draft of your document, including rough versions of visual elements and design. Test your draft on 4–5 potential readers who are participants in the organization for which you are creating the document. Take care to choose readers who represent a range of knowledge about the topic, familiarity with the genre and format chosen, and the cultural, age, and geographic distribution of the members of the organization. Ask each of them to read the document, paraphrase it in their own words, explain the purpose of the document, answer 2–3 questions on its content, and describe their reaction to the tone and design of the document. After you have discussed the document with each reader, bring all of them together in a focus group to discuss their responses and to offer suggestions for revision.

Write a **report** in which you explain the details of how you conducted your test and summarize the results, including the suggestions for revision. What aspects of your document need to be changed and how? What further background information do you need to explain? What differences in needs among the readers will you need to address in your revision?

Write a **reflection** in which you discuss what you learned from conducting the test. What kinds of information did you get from readers that you did not get from your research? What problems did you have in conducting the test? What strategies would you use to make such testing work better in the future?

Stage 4: Testing Second Draft

Write a second draft of your document, including rough versions of visual elements and design. Pay attention to all aspects of the document, including organization, design, and visuals. Consider what strategies you can use to address different readers' contrasting needs, paying particular attention to strategies that enable readers to fill in their own contextually dependent interpretations and to strategies that evoke readers' Background skills. Include in the introduction a brief description of the purpose of the document and the context within which it is designed to be used. Also incorporate in this draft what you have learned about strategies for clear communication in class, being careful to consider the needs of your particular readers and the context and purpose of the communication as you do so.

Recruit 4–5 new potential readers (not the same ones you used before). Devise a series of tasks or questions that enable you to see if readers can use the document as it is intended. Observe each reader, asking them to talk about their thinking as they work on the tasks or questions. Take careful notes on what they say, or record their comments.

Write a **report** in which you explain the details of how you conducted your test and summarize the results. Based on what you have learned from this test, what aspects of your document need to be changed and how?

Write a **reflection** in which you discuss what you learned from conducting the test. What additional information did you learn from this test that was not obvious in the test of the first draft? What problems did you have in conducting the test? What strategies would you use to make such testing work better in the future?

Stage 5: Testing Final Draft

Create a final draft of the document, making all aspects, including the design and visuals, as complete and professional looking as you can. Make at least three copies of the final draft.

Take the copies of the draft to the site where readers will be using it. Talk with 3–4 organization participants about its purpose and what you hope it will help them do. Then ask them to use it in the context in which it was designed to be used.

Observe them using it, and ask them to talk about their responses. Then ask them to discuss with you their reactions to and evaluation of the document.

Write a **report** in which you discuss what you learned. How well did your document work with its readers? What aspects of the context of use did you take enough account of, and what aspects did you not account for in your document? What would you change if you were to do another draft of the document?

Stage 6: Commentary

Look over all your drafts, reports, and reflections, and write a critical commentary on all your work on the project. What strategies for working with readers did you learn? What would you do differently? Referring also to what you have learned in class and from the class readings about writing for readers, explain what principles enable a writer to effectively attend to readers' needs in documents.

REFERENCES

Botka, D. H. (2002). From gobbledegook to plain English: How a large state agency took on the bureaucratic form letter. *Proceedings of the STC Conference.*

Bushnell, J. (1999). A contrary view of the technical writing classroom: Notes toward future discussion. *Technical Communication Quarterly, 8,* 175-188.

Cooper, M. M. (1996). The postmodern space of operator's manuals. *Technical Communication Quarterly, 5,* 385-410.

Dobrin, D. A. (1989). *Writing and technique.* Urbana, IL: NCTE.

e-LearningHub.com. (2002). A systematic approach to instructional design. http://www.e-learninghub.com/articles/systematic_instructional_design.html. Retrieved April 1, 2005.

Giles, T. D., & Still, B. (2005). A syntactic approach to readability. *Journal of Technical Writing and Communication, 35,* 47-70.

Glanz, J. (1997). Fermilab group tries plain English. *Science, 276.* http://www.sciencemag.org. Retrieved March 2004.

Houp, K. W., Pearsall, T. E., Tebeaux, E., & Dragga, S. (2002). *Reporting technical information* (10th ed.). New York: Oxford University Press.

Kimble, J. (1996-97). Writing for dollars, writing to please. *Scribes Journal of Legal Writing, 6,* 1-38.

Medvedev, P. N., & Bakhtin, M. M. (1978). *The formal method in literary scholarship: A critical introduction to sociological poetics.* A. K. Wehrle (Trans.). Baltimore, MD: Johns Hopkins University Press.

Miller, C. (1979). A humanistic rationale for technical writing. *College English, 40,* 610-617.

Mirel, B. (1998). "Applied constructivism" for user documentation: Alternatives to conventional task orientation. *Journal of Business and Technical Communication, 12,* 7-49.

National Commission on Writing for America's Families, Schools, and Colleges. (2004). *Writing: A ticket to work or a ticket out . . ..: A survey of business leaders.* College Entrance Examination Board. Available online at http://www.collegeboard.com.

A plain English handbook: how to create clear SEC disclosure documents. (1998). Office of Investor Education and Assistance, U.S. Security and Exchange Commission. Available online at http://www.sec.gov.

Schneider, B. (2002). Clarity in context: Rethinking misunderstanding. *Technical Communication, 49,* 210-218.

Tyler, K. (2003). Toning up communications. *HR Magazine,* 48. http://www.shrm.org/hrmagazine/articles/0303/0303agn-training.asp. Retrieved April 1, 2005.

Van Horen, F. M., Jansen, C., Maes, A., Noordman, L. G. M. (2001). Manuals for the elderly: Which information cannot be missed? *Journal of Technical Writing and Communication, 31,* 415-431.

Weiss, T. (1993). "The gods must be crazy": The challenge of the intercultural. *Journal of Business and Technical Communication, 7,* 196-217.

Weiss, T. (1997). Reading culture: Professional communication as translation. *Journal of Business and Technical Communication, 11,* 321-338.

Contributors

GARY BAYS has taught composition and technical communication at the University of Akron Wayne College for 20 years. He was the first, if unofficial, graduate student in the Humanities Department at Michigan Technological University (thanks to a joint teaching internship with Central Michigan University). Bays served on the editorial board of the *Journal of the Conference on College Composition* and Communication and is the editor of the *Ohio Association of Two-Year Colleges (OATYC) Journal*. His range of scholarship includes workplace communication, scholarship in the two-year college, and collaborative learning. Most recently, his chapter on workplace research was included in *Innovative Approaches to Technical Communication Pedagogy* (Utah State University Press, 2004).

TRACY BRIDGEFORD is Associate Professor at the University of Nebraska at Omaha, where she directs the Graduate Certificate in Technical Communication. She coedited (with Karla Kitalong & Dickie Selfe) *Innovative Approaches to Technical Communication Pedagogy* (Utah State University Press, 2004). Her chapter in this collection, "Story Time: Teaching Technical Communication as a Narrative Way of Knowing," describes the role of narrative in her technical communication courses. She is currently working on a book proposal about digital literacies.

MARILYN M. COOPER is Professor of Humanities at Michigan Technological University and past editor of College Composition and Communication. She approaches technical communication through the lens of theories of language and meaning and is the author of "The Postmodern Space of Operator's Manuals" (*Technical Communication Quarterly,* 1996).

DÀNIELLE NICOLE DeVOSS is Associate Professor and Director of the Professional Writing Program at Michigan State University. Her research interests include computer/technological literacies, feminist interpretations of and interventions in computer technologies, philosophy of technology/technoscience, professional writing, technical communication, gender/identity play in online spaces, online representation and embodiment, and issues of rhetoric in disciplines such as nursing and medicine. DeVoss's work has most recently appeared in *Computers and Composition; Journal of Business and Technical Communication; Pedagogy: Critical Approaches to Teaching Literature, Language, Composition, and Culture; Moving a Mountain: Transforming the Role of Contingent Faculty in Composition*

Studies and Higher Education (2001); and *Writing Center Research: Extending the Conversation* (2001). She recently coedited a collection on behavioral interventions in cancer care, *Evidence-based Cancer Care and Prevention* (2003), and the collection *Digital Writing Research: Technologies, Methodologies, and Ethical Issues* (with Heidi McKee, forthcoming).

PATRICIA FREITAG ERICSSON is Assistant Professor in the English Department at Washington State University, where she teaches undergraduate and graduate courses in technical and professional communication and digital technology and culture. Her research interests include technical writing, techno-rhetoric, and the intersections of technology, education, and agency theory. Previously, she taught and directed the writing program at Dakota State University. Her work has appeared in a variety of venues, including *Computers and Composition; The ACE Journal; Text Technology; English Education;* and several edited collections. She is coediting the volume *Machine Scoring of Student Essays: Truth and Consequences* (Utah State University Press, forthcoming).

JOHNDAN JOHNSON-EILOLA is Professor of Communication and Media at Clarkson University, where he teaches courses in information architecture, mass media, new media, and rhetoric. His recent books include *Professional Writing Online* (Allyn & Bacon/Longman, 2002), *Central Works in Technical Communication* (with Stuart A. Selber; Oxford University Press, 2004), *Writing New Media* (with Anne Wysocki, Cynthia Selfe, & Geoffrey Sirc; Utah University Press, 2004), and *Datacloud* (Hampton Press, 2005).

JAMES KALMBACH is Professor of English at Illinois State University, where he teaches courses in computer literacy, English education, digital rhetoric, and technical communication. He is the author of *The Computer and the Page: Publishing, Technology, and the Classroom* (Ablex, 1997) and coeditor of "Beyond Normal: Teaching and Learning in Virtual Spaces" (http://english.ttu.edu/kairos/7.3/index/html), a collection of papers from *Computers and Writing* (2002). His current research focuses on new media, Web authoring, and the study of emergence/improvisational pedagogies in technology-rich classrooms.

KARLA SAARI KITALONG left Michigan Technological University in 1999 to join the faculty at the University of Central Florida in Orlando, where she is Associate Professor in the university's Bachelor's and Master's Programs in Technical Communication and the Ph.D. in Texts and Technology. Her research and teaching interests center on visual communication and usability. Recent publications include Innovative Approaches to Technical Communication Pedagogy (coedited with Dickie Selfe & Tracy Bridgeford; Utah State University Press, 2004); chapters on the cultural presence of image editing software and practices and on the invisibility of teachers in online learning; and a classroom supplement on visual communication. Kitalong often teaches fully online courses; as a result, she developed and teaches a doctoral course in online writing pedagogy and mentors graduate teaching assistants as they teach their first online writing courses.

ANN KITALONG-WILL is Adjunct English Instructor at Alpena Community College, where she teaches technical communication and journalism classes. She also works as In-home Educator with Northeast Michigan Even Start, a family literacy program that assists parents in realizing their roles as their child's first and most important teacher, in conjunction with meeting their own educational goals. She works with at-risk families and teen parents who wish to complete their high school diplomas or GED certificates, and provides interactive literacy activities for parents and children to learn how to learn together.

TERESA KYNELL HUNT is Interim Assistant Vice President for Academic Affairs at Northern Michigan University. She holds a Ph.D. in rhetoric/technical communication from Michigan Technological University. Past chair of the NCTE Committee on Technical and Scientific Communication and Executive Committee member of the ATTW, she has authored a variety of articles, including the Nell Ann Pickett Award-winning "Technical Communication from 1850-1950: Where Have We Been?" She is the author of *Writing in a Milieu of Utility;* coeditor (with Michael Moran) of *Three Keys to the Past: The History of Technical Communication;* coeditor (with Gerald Savage) of *Power and Legitimacy in Technical Communication,* Vols. I and II; and coauthor (with Wendy Stone) of *Scenarios for Technical Communication: Critical Thinking and Writing.* Kynell Hunt is a member of the Peer Review Corps of the Higher Learning Commission and consults externally on assessment and accreditation.

MICHAEL MARTIN received his Ph.D. from Michigan Technological University and is Assistant Professor at the University of Washington–Stout, where he teaches both graduate and undergraduate students in technical writing and writing technical manuals. He also teaches hypertext writing and business writing. In a cross-discipline collaboration, he has developed and is teaching a business writing course specifically for hospitality, restaurant management, and tourism majors. Formerly her wrote manuals and did corporate training for Daimler Chrysler and Gateway Computers. Martin also current coordinates and supervises co-ops and internships for the technical communication program at Stout. His work focuses on ethics and usability in the classroom and in the workplace, and he integrates ethics and rhetorical strategies for teaching ethics in the courses he teaches.

MICHAEL R. MOORE teaches at Michigan Technological University in the Department of Humanities and in the College of Engineering and does research in literacy, composition, and technical communication. He has published in professional journals and presented papers at national conferences in the areas of computer-mediated pedagogy, teacher education, and research on writing and literacy practices. His most recent scholarly work has been in developing alternative textual and contextual analytic frameworks for use in English studies, and he is currently working on a history of the institutionalization of literacy research.

PETER PRAETORIUS teaches technical writing and communication courses at Matanuska-Susitna College in Palmer, Alaska. He is the author of "Technical Communicators as Purveyors of Common Sense" (*Journal of Technical Writing and Communication*). He received his bachelor's in biology from Whittier College, his master's in professional communication from Clemson University, and his Ph.D. in rhetoric and technical communication from Michigan Technological University.

GERALD J. SAVAGE holds a Ph.D. in rhetoric and technical communication from Michigan Technological University. He teaches technical communication, rhetoric, and editing courses and directs the technical writing program and the English studies internship program at Illinois State University. His research includes studies of engineering writing, professionalization issues in technical communication, technical communication ethics, and technical communication literacy practices. Savage has worked as a technical writer and editor in the software industry, education, and government. He coedited (with Dale Sullivan) *Writing a Professional Life: Stories of Technical Communicators On and Off the Job* and (with Teresa Kynell Hunt) *Power and Legitimacy in Technical Communication,* Vols. I and II.

STUART SELBER is Associate Professor of English at Penn State and a past president of the Council for Programs in Technical and Scientific Communication. He is the author of Multiliteracies for a Digital Age (Southern Illinois University Press, 2004) and coeditor (with Johndan Johnson-Eilola) of Central Works in Technical Communication (Oxford University Press, 2004). Both books have received a publication award.

RICHARD (DICKIE) SELFE consults across the College of Humanities on instructional technology projects and designs support systems for teachers at The Ohio State University. His interests lie at the intersection of communication pedagogies, programmatic curricula, and the social/institutional influences of digital systems. His most recent book-length project is *Sustainable Communication Practices: Creating a Culture of Support for Technology-rich Education.* Selfe's recent publications also include "Teacher Quality: The Perspective of NCTE Members" (*English Education,* forthcoming) and, as author/coeditor, *Innovative Approaches to Technical Communication Pedagogy* (Utah State University Press, 2004).

JENNIFER SHEPPARD is Assistant Professor of Rhetoric and Professional Communication at New Mexico State University, where she teaches courses in technical and scientific communication, multimedia theory and production, Web development, and document design. Her research focuses on the intersections of digital multimedia development, workplace communication, and literacy theory. She has conducted qualitative studies investigating these interests in a number of classroom and professional settings. This research has included work with the U.S. Forest Service, for which Sheppard created and studied the development of an interactive, science-based multimedia Web site for middle-school students based on ecological field research.

SUMMER SMITH TAYLOR is Assistant Professor and Director of the Masters in Professional Communication program in the English Department at Clemson University. Her research focuses on the teaching and evaluation of writing, particularly in engineering. She teaches technical writing, research methodologies, and the teaching of technical writing. As Director of the Advanced Writing Program at Clemson, she developed a client-based program for undergraduate technical writing classes. The client-based program was cited as one of the main reasons the Advanced Writing Program was honored by the Conference on College Composition and Communication with the Writing Program Certificate of Excellence in 2004.

ANNE FRANCES WYSOCKI is Associate Professor of Visual and Digital Communication in the Humanities Department at Michigan Technological University, where she teaches courses in visual and verbal composition, visual rhetoric, and new media. She is lead author of *Writing New Media: Theory and Applications for Expanding the Teaching of Composition,* which won the 2005 Computers and Writing Distinguished Book Award, and her compositions have appeared in *Computers and Composition, Kairos,* and the *Journal of the Council of Writing Program Administrators,* as well as in many books. Her interactive new media piece, "Leaved Life," was a winner of the Institute for the Future of the Book's 2005 Born Digital Competition. Wysocki has done multimedia, instructional, and curriculum design for such clients as Apple, the Los Angeles Conservations Corps, Microsoft, the Corporation for National Service, and Public/Private Ventures.

ART YOUNG is Robert S. Campbell Chair in Technical Communication, Professor of English, and Professor of Engineering at Clemson University. He is the founder and coordinator of Clemson's communication-across-the-curriculum program, a university-wide initiative to improve the communication abilities of all Clemson students. In March 2002, he received the Exemplar Award from the Conference on College Composition and Communication for outstanding achievement in teaching, research, and service. He has served as a consultant on communication across the curriculum, technical communication, and program evaluation to more than 70 colleges in the United States and abroad.

Index